AF553424

UNDERSTANDING NANOTECHNOLOGY

ENCYCLOPAEDIA OF NANOTECHNOLOGY - I

UNDERSTANDING NANOTECHNOLOGY

By

Dr. M.P. Arora

Dept. of Zoology
M.M.H. Post Graduate College
Ghaziabad
(U.P.)

DISCOVERY
PUBLISHING HOUSE

First Published – 2008

Reprinted – 2025

ISBN: 978-93-5056-581-0 (Set)
978-81-8356-277-5

Understanding Nanotechnology

Published by:

DISCOVERY PUBLISHING HOUSE
4383/4B, Ansari Road, Darya Ganj
New Delhi-110 002 (India)
Phone: +91-11-23279245; 23253475; 43596065
Mobile: +91 9811179893 / +91 9871656464
E-mail: discoverybooksindia@gmail.com
orderdphbooks@gmail.com
namitwasan9@gmail.com
web: www.discoverypublishinggroup.com

Printed at:
Infinity Imaging Systems
Delhi

Preface

Everyone in the modern world knows what technology is. But what is nanotechnology? Taken from the Greek, *nano* means "one billionth part of" a whole. In modern parlance, it means very, very small. In today's world of smaller cell phones and portable computers, miniaturization is the current technology. Nanotechnology is the next step after miniaturization. In tomorrow's world, nanotechnology will be the new common technology. It will affect everyone of the planet and many change civilization as we know it.

Nanotechnology is the molecular engineering. It will soon create effective machines as small as DNA. It describes the ideas and techniques that are creating or new domain of science and technology. Nanotechnology holds the promise of revolutionizing field ranging from medical science to optical communications, but requires a substantial investment in time and money to make the promise become a readily. The results could be nothing less than a new industrial revolution.

In an undertaking as new as nanotechnology, many different approaches are studied in order to ascertain the best way to fashion working devices. At present, techniques can be divided into one of two general categories. One technique, termed the top-down method, physically manipulates molecules into nanostructures. The other technique, termed the bottom-up method, coaxes atoms and molecules to assemble themselves into nanostructures.

The present title **"Understanding Nanotechnology"** has been designed for undergraduate and post-graduate students as well as those involved in basic research in biological, biochemical and biophysical sciences.

In the preparation of this book large number of books and research papers have been consulted. So no authenticity is claimed.

The author expresses his gratitude to Mr. Wasan and staff of M/s Discovery Publishing House for their whole hearted co-operation in the publication of this book.

The author tried hard to be accurate and upto date in statement and realises the impossibility of completely avoiding errors therefore, the author will greatly appreciate having his attention called to any questionable statement.

Author

CONTENTS

1

INTRODUCTION

The first half of the twentieth century witnessed an explosion of technology that deeply affected the way we live. In 1900, heavier-than-*air flying machines* were widely believed impossible; in 1950, jets were approaching the speed of sound. In 1900, most people did not have cars, electricity, or indoor plumbing. By 1950, they did. The same fifty years saw substantial parts of the development of *antibiotics*, radio, television, plastics, *nuclear weapons*, and the *computer*. Tractors, harvesters, and similar equipment cut the number of people required to produce a given quantity of food by a factor of ten.

To a great extent, the march of progress continued through the second half of the century. *Jet airliners* became common, and with the Boeing 747, air freight became economical for some kinds of goods. Televisions, computers, cell phones, and similar gadgetry became ubiquitous. The global communications network along with the construction of enormous freight ships and tankers wrought a world economy more integrated at the turn of the twenty-first century than the national one had been at the turn of the twentieth. Men walked on the Moon.

And yet somehow the grand promise of technology seemed to lose its magic. The footprints on the Moon have lain undisturbed for decades. In the late twentieth century, Western civilization produced an artifact with a volume surpassing the Great Wall of China. It was the Fresh Kills landfill, New York City garbage dump on southern Staten Island. Automobiles do not go very fast when the roads are jammed full of them. Even the production of food in unheard-of quantities resulted in an epidemic of obesity and heart disease.

The shift in perceptions is complex but has a few main roots. First is simply human nature: the promise of technology in the early twentieth century was largely fulfilled, in the industrialized nations at least. People who are well fed, warm, and not faced with hard physical labour turn their attention elsewhere. We did not have two global wars in the second half century as we did in the first. War has a tendency to focus attention toward the means of victory, and away from undesirable side effects.

If you had told a farmer in 1900 that his descendants a century thence would spend their time mostly inside heated and air-conditioned (air what?) buildings, sitting on cushioned seats, talking, reading, and writing, he might have believed you if he was a believer in Progress. If you had told him they would call this "working," he might have laughed in your face. The triumphs of twentieth-century technology—full stomachs and the lack of having to do backbreaking labour—are taken for granted, and lesser problems have been elevated in their stead.

And finally, there is a real phenomenon of diminishing returns. A safety razor with two blades is not twice as useful as one with only one blade. Cars hit a speed limit, and thus a usefulness limit, because of the limitations of human reflexes, not limitations of mechanical capability. Airliners hit a speed limit involving economics and optimal flight regimes. The technology for *supersonic passenger flight* is there, it just costs too much to pay its way.

In science, the first half of the twentieth century saw a vast flowering of knowledge in chemistry, as physics revolutionized the way we understand ordinary matter with quantum mechanics. It's reasonable to say that a scientist in 1950 could give a good, fundamentally sound explanation of ordinary situations, such as why materials have the properties they do. The scientist of 1900 would have had to do a lot of fudging. He could have told you that snowflakes have six fold symmetry because the water molecule has a certain shape. The scientist of 1950 could tell you why the molecule has that shape.

What about the latter half of the century? Did science hit the same apparent law of diminishing returns as did technology? In some areas, yes. Physics, after its brilliant success, moved on to mostly more esoteric phenomena, having less to do with the everyday world. The fact that *protons*, *neutrons*, and so forth are made from quarks is a marvelous systematization of subnuclear physics, but it has no practical impact on anyone's life. But something else did. Without nearly as much fanfare as it deserved, science as a whole proceeded to crack open the greatest, most essential mystery of all time: the nature and mechanism of life itself.

The story really begins in the early 1600s. In 1633, around the time Galileo was being condemned by the Church., Rene Descartes wrote a book called *De Homine* in which he tried to explain some of the phenomena of the human body in mechanistic terms. Descartes is famous for the *doctrine of Dualism*, the claim that there is a distinction between mind, which acts mysteriously, and matter, which works mechanically. *Vitalism*, belief in a "*life force*" that does not obey the mechanistic laws of physics and chemistry, had a strong history in intellectual tradition and an even stronger presence in peoples everyday understanding of how things worked. Descartes stuck a foot in the door. He opened the way to seek mechanistic explanations for the phenomena of life. Dualism, intentionally or not, formed the perfect shield against the reactionary and religious forces of the day, allowing one to seek mechanisms for hand—eye coordination, for example, while leaving spirituality; morality, and consciousness to the category of "mind."

Fast-forward to 1825. Chemists were still discovering new elements—the basic rules and properties of how they combine are still being worked out—and they divided the world into two distinct kinds of "*stuff*": inorganic and organic. Never the twain shall meet. Organic compounds were made by living things, while inorganic ones could be made by chemists in test tubes. You could break organic compounds down, of course; just burn them, for example. But putting them together required some ineffable life force that was universally and seriously believed to exist. Then a young chemistry teacher named Friedrich Wohler produced an inert solid from a mixture of aqueous hydrogen cyanide and ammonia in 1825, subsequently making the same chemical in different ways from cyanide and ammonia salts. By 1828 he had managed to show that his compound was chemically identical to urea, an organic compound found in urine.

By the twentieth century, progress in everything from physiology to chemistry had made it reasonable to think that all of life was basically mechanistic. However, it was still largely a matter of faith either way. The mechanisms were not known. By 1944, though, things had advanced to the point that Erwin Schrodinger could write *What Is Life?* This landmark essay made a persuasive case that all the phenomena of life at the cellular level could be part of an unbroken chain of explanation going all the way down to quantum mechanics (of which Schrodinger was one of the main developers). Nine years later, Francis Crick and James Watson discovered the *structure of DNA*, and the floodgates opened.

Once upon a time when in a grammar school, biology teacher used to explain four properties made living things different from nonliving ones. These were Organization, Metabolism, Reproduction, and Irritability; Organization meant that organisms have complex inner structure, being made of cells, and cells having a further inner structure. Metabolism meant that they consumed nourishment for growth and activity. Reproduction meant—well, this was a grammar school. And Irritability meant reactivity to stimuli ranging from the conscious actions of humans to the tropisms of plants.

What the biology teacher did not explain, partly because it was an introductory course and partly because it wasn't fully understood at the time, was how all those properties are the result of the activities of molecular machines. The key is reproduction. Cells reproduce to form the structure of the body, and ultimately to allow the entire organism to *reproduce*. A *cell* is a factory filled with machines and a blueprint. Among the machines are machines to make the machines, called *ribosomes*, and machines to copy the blueprint, called *replisomes*. The blueprint itself comes in volumes called *chromosomes*. Each one is a molecule of *DNA*.

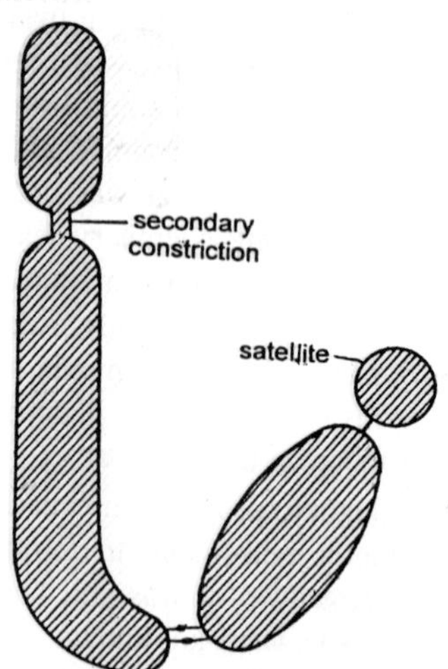

Fig. 1.1. Metacentric chromosome.

In the latter half of the twentieth century; we went from having a sketchy notion of how it might work to having our teeth solidly into the details. The "*genetic code*," deciphered in the 1960s, tells what sequence of amino acids a given piece of DNA is the blueprint for. We cannot say, in general, what protein shape that sequence will fold up into, nor the function of the molecular machine the protein forms, but we are getting there. We know a lot of specific cases. A gene is a part of the DNA that's the description of a given machine. We can snip genes from the DNA of one kind of organism and insert them into another, with various results, including glow-in-the-dark pet fish.

All the forms of life we know contain many molecular machines. The ability to tinker with them, put them together in new combinations, and modify life in various ways, is called *biotechnology*. As yet, biotechnology has not done much in the way of brand-new, invented designs of molecular machines, but that's coming. Biotechnology is limited, however. To begin with, all the machines are proteins. (Biology can build with other materials, creating bone and teeth, but not active molecular machines.) It's stuck with the DNA—RNA—ribosomes some amino acid—protein mechanism, and that means it's limited to operating in solution in a narrow temperature range. There are plenty is things it cannot make: tin cans, for example.

On the other hand, biology—and biotechnology when we master it—can and does make a staggering variety of mind-bogglingly complex, incredibly adaptable things—creatures—with capabilities our conventional manufacturing technology cannot match. We can build Moon rockets, which biology cannot, but not insects, which biology does with mad abandon. We can build *electron microscopes*, which biology cannot, but Joyce Kilmer's dictum that only God can make a tree is still true. A nuclear-powered aircraft carrier, loaded with planes, has something like 100 trillion working parts (most of which are the trench capacitors that store bits in the memories of its computers). So does a housefly.

What if it were possible to combine the capabilities of the biological mechanisms with those of conventional ones? We could have machines that could grow and repair themselves, proliferate without factories, grow from seeds. We could have a technology with the subtlety and adaptability of life, the power of jumbo jets, the efficiency of electric motors, the precision of computers. Is such a technology possible?

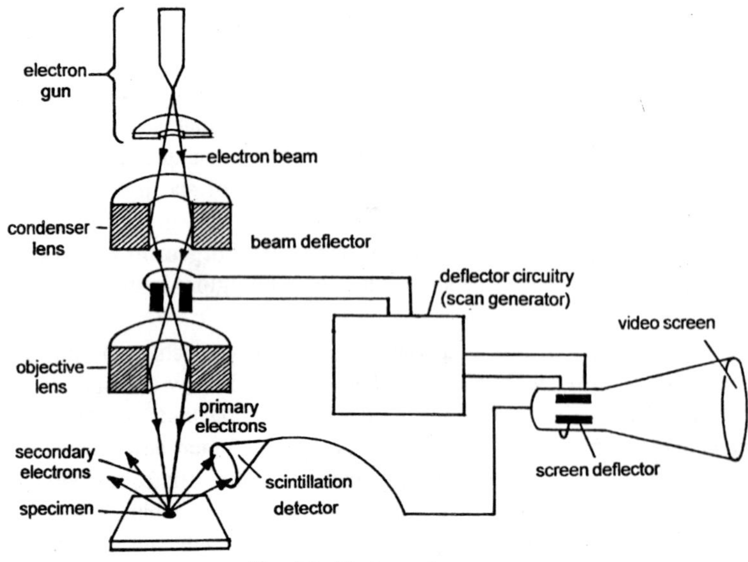

Fig. 1.2. Electron microscope.

It's not only possible, it's on its way. It will change the way we live, what we know, how we think, and maybe even who we are. It is called nanotechnology.

What is Nanotechnology?

Nanotechnology: the art of manipulating materials on an atomic or molecular scale especially to build microscopic devices (as robots).

Most books about future technology tend to follow Clarke's Law: "Any sufficiently advanced technology is indistinguishable from magic." It is difficult to do otherwise: for the general reader, and indeed to the specialist outside his field, much of existing technology is magic. This is not, however, a very useful basis for the reader who wishes to make personal long—term plans, participate in the political and economic decision-making processes, or simply have a feeling of understanding how things work in the world.

If everyone commenting in public about *nanotechnology* were saying basically the same thing, it would be at least reasonable just to take their word for it. However, there seems to be no widespread agreement on what nanotechnology is, much less what it can do.

One of the first places that the term got wide play was in science fiction. This at least let some of the ideas be discussed, but science fiction is not necessarily a good place to learn about new and important technological developments. In the first place, the science fiction writer is under no obligation to stick to the known, the possible, or even the truth. The demands of telling a good, dramatic story come first. So the science fiction writer will often write the plot first and then cut, stretch, or simply make up the technology to fit.

Here are a few of the things nanotechnology is not: it is not infestations of "*nanites*" that take over your starship, as in *Star Trek*; it is not evil shape—shifting killer robots from the future, as in *Terminator II*; it is not evil clouds of flesh—eating mites, as in *Prey*. These things are science fiction, with the emphasis on fiction. They are designed by storytellers to have a maximum emotional impact, and have virtually nothing to do with sensible, levelheaded technological prediction.

On the other hand, you'll find people advertising the most mundane advances as nanotechnology: stain-resistant pants, skin creams, oil and paint additives, and so forth, which do not seem significantly different from other cloth treatments, additives, and what have you. Indeed, as write this, New York's attorney general has been asked to investigate the overuse of the word *nanotechnology* in company descriptions over-promotion. One commentator quipped, "It seems that they're willing to call themselves 'nanotechnology' if their product is made of atoms."

It may help to have a took at the history of the word *nanotechnology* itself. Back in the 1970s, biotechnology was beginning to emerge as a serious possibility with the introduction of recombinant DNA techniques. K. Eric Drexler, an MIT student, became interested in the notion that much of what went on inside a cell was not so different, in principle, from what engineers did at macroscopic scales. What's more it seemed reasonable that, having gained control of the cell's molecular machinery, one could use it the same way that engineers did normal-size machines: making materials, structures, tools, and more machines.

Drexler wrote these ideas in a scientific paper and published it in the *Proceedings of the National Academy of Sciences* in 1981, under the title, "*Molecular Engineering*: An Approach to the Development of general Capabilities for Molecular Manipulation." Its abstract was as follow:

Development of the ability to design protein molecules will open a path to the fabrication of devices to complex atomic specifications, thus sidestepping obstacles facing conventional *microtechnology*. I path will involve construction of molecular machinery able to position reactive groups to atomic precision. It could lead to great advances in computational devices and in the ability to manipulate biological materials. The existence of this path has implications for the present.

During the 1980s, Drexler and a group of friends worked out the possibilities and some of the implications of such a technology. Then in 1987, Drexler published a popular book about it entitled *Engines of Creation: The Coming Era of Nanotechnology*. The people who had been thinking about it and working on it throughout the 1980s had used the word *nanotechnology* all along; but it was the publication of *Engines* that first introduced the term to the public at large.'

Engines of Creation is a technophile's dreamscape. It predicts microscopic replicating units able to build skyscraper-sized objects to atomic precision. These could be buildings or they could be spaceships. It discusses artificial intelligence and engineering systems able to handle the enormous complexity such designs would require. It speaks of "*easy and convenient*" space travel, and describes a spacesuit so light and thin that you almost forget you're wearing it (present-day spacesuits are very awkward and arduous to wear and work in). It mentions cell repair machines and curing "a disease called *aging*." It talks about cryonics and resurrecting the frozen.

Engines created a sensation in technically oriented circles. There had always been a segment of the population who liked technology for its own sake, or believed it could, and dreamed it would, continually improve the human condition. (This segment is actually smaller now than it was a century ago, when Edison was a popular hero.) These dreams now had a form, with pathways and techniques in sight where only the vague mantra of "technological progress" had been before. They also had a name.

Drexler coined the word *nanotechnology* by analogy to the then already common *microtechnology*, which mostly applied to photo-lithographic chip-making techniques but was broadly applied to any technology that manipulated matter at the micron scale. Since the new technology would manipulate matter at the nanometer scale, the extension seemed straightforward.

By the mid-1990s, *nanotechnology* had picked up the cachet of a popular buzzword, and started showing up in science fiction, popular scientific, and technical publications—and grant proposals. After all, if *nanotechnology* means dealing with matter on the scale of nanometers, then a huge amount of existing science and engineering was, by definition, *nanotechnology*: chemistry, molecular biology surface physics, thin films, ultra fine powders, and so forth. By the turn of the century, computer chips, the original *microtechnology*, had features small enough that they could he reasonably measured in nanometers. (A micron is 1,000 nanometers, so instead of saying 0.08 microns you say 80 nanometers.)

Popularity and "Success"

In a perfect world, a definition would have been an issue for lexicographers. But in reality, the world of science, no less than the world of politics, is driven by fads and buzzwords. And where the two intersect-the science and technology funding agencies you need a buzz saw to get through the buzzwords. So the word *nanotechnology* was rapidly adopted by researchers attempting to get just that one extra edge over their peers in the highly competitive, not to say cutthroat, business of getting government money.

Another thing that funded researchers compete for is the brightest graduate students. Graduate students actually do most of the work in university labs, and for wages that are quite low compared with the standards of the commercial world. Researchers offer a number of non monetary rewards, and one of the main ones is for the students to be working in, and learning, new fields at the cutting edge knowledge. Describing your work as nanotechnology was a good way to make it more attractive to the best students.

So the meaning got stretched, and stretched, and stretched. And then a funny thing happened. By the turn of the century, there was a *National Nanotechnology Initiative*, and roughly a billion dollars per year of funding for research under the name. And as will happen, people who wanted the money for other things, and people who just didn't like science and technology in the first place, and journalists looking for an eye-catching story, all went back and read Drexler again and the first thing they found was the chapter entitled "*Engines of Destruction.*"

Of course, any powerful new technology will bring new dangers; and nanotechnology being essentially engineering at the level of control that gives life its miraculous powers, is very powerful indeed. So the technophobes, Luddites, and horror fiction writers had a new Frankenstein's monster to play with.

This left the "*nanotechnology stretched version*" researchers in an uncomfortable, if ironic, situation. They had adopted a name for their endeavors that was a popular buzzword because of the association it had with powerful future technologies and revolutionary capabilities. But what they were working on was really just more chemistry, or more ultra fine powders, or what have you. And now they were being attacked for doing demon's spawn that could destroy everything if the slightest mistake were made. Their immediate, and quite predictable reaction: "Oh, no, *nanotechnology* cannot do that kind of thing."

And they are quite right; the stuff that's going on in most labs today under the name of *nanotechnology* may make smaller computer chips, or stronger aerospace materials, or whatever, but it is really more of the same old conventional technology by another name. You do not need to read a whole new book to learn that people are trying to make more stain-resistant (and expensive) pants, or stronger (and more expensive) tennis racquets, or smaller, faster computers. Nor do you need to worry over the fact that marketing departments will be calling these things, and lots of other things over the coming years, "*nanotechnology*." It is just a word.

There is the appearance of a debate going on in the scientific community as to whether original, Drexlerian *nanotechnology* could work as advertised. The appearance is deceiving. There are indeed two camps of thought: one is the politically motivated people, largely grant funded, who are worried about the public perception of nanotech dangers impacting their money. They have very much a "not invented here" attitude toward anything even vaguely smelling of self- replicating machines or molecular manufacturing, and have made little effort to understand it. Typically, though not always, you'll find chemists and materials scientists on this side of the question.

The other side is not so politically oriented, does not have the money to protect in places where it is vulnerable to every passing wind of public opinion, and—surprise—is not nearly as skeptical. A substantial proportion of scientists and engineers, those without a monetary stake in the technical question, agree with the position I take in this book. Typically, but again not always, you will find physicists, mechanical engineers, and computer scientists on this side.

Drexler did his doctoral dissertation at MIT on *nanotechnology*, and then after some further research published it as a technical book entitled *Nanosystems*: *Molecular Machinery*, *Manufacturing*, and *Computation*. No one has ever found a significant error in the technical argument Drexler's detractors in the political argument do not even talk about it; they tend to avoid technical argument altogether, resorting instead to emotional non sequiturs like, "You are frightening our children."

Some of the potentialities of nanotechnology are indeed frightening. But that makes it even more important to think about it seriously—not to avoid it! It's my considered opinion, after more than a decade of technical study, that nanotechnology as Drexler described it is almost certainly feasible, and nearly as certain to be implemented by somebody, somewhere, within the current century. Given its profound implications for the human condition, it would be incredibly foolish not to weigh it carefully in our planning.

WHAT IS NANOTECHNOLOGY?

But even in the field of science, it is perilous to run counter to the accepted tables of precedence. On no account is it permissible to mention living beings and machines in the same breath. Living beings are living beings in all their parts; while machines are made of metals and other unorganized substances, with no fine structure relevant to their purposive or quasi-purposive function.

So *nanotechnology* really does have two different meanings. One is the board, stretched version meaning any technology dealing with some thing less than 100 nanometers in size. The other is the original meaning: designing and building machines in which every atom and chemical bond is specified precisely. The capabilities and dangers of nanoscale technology are simple and straightforward extensions of current trends in the capabilities and dangers of chemistry; materials science, and microfabrication. The majority of new techniques being discovered and trumpeted as the "latest thing in nanotechnology" today will be obsolete in ten years.

Where a term is needed to make the distinction, the best word seems to be *eutactic*. Eutactic means "well ordered" and has the same import in the context of nanotechnology as the phrases *atomically precise* or *low entropy*. Specifically, it means that in the system being talked about, there is a place for every atom, and every atom is in its place. Eutaxy is not necessarily nanoscale: a crystal is precisely ordered at the atomic scale, yet can be of macroscopic size. Turbine vanes in modern jet engines are grown as single crystals for strength at high temperatures. Quartz is eutactic but glass is not.

So what, then, is (eutactic) nanotechnology? It is a technology that does not physically exist today, but can be straightforwardly analyzed, modeled, and simulated based on very standard well-understood science and engineering. It involves building machines whose parts are of molecular size, but more important of atomic precision: each atom and bond in the finished part is called for specially in its design, just as the parts in the machinery of the cell are. In a mature nanotechnology, however, they will operate in a vacuum instead of salty water, and will be made of materials much stiffer and stronger than the proteins that cells use.

Given that, we can simply copy existing mechanical designs of assembly lines down to the molecular scale, and build as wide a variety of machines there as we now do at everyday scales. That's just a start, of course—many things will work differently and need to be redesigned. Some things will not work as well (centrifugal pumps, for example) but others that do not work well at macroscales will work at the nanoscale (electrostatic motors, for example).

Perhaps surprisingly, we can say more about such an ultimate technology than we can about all the different ad hoc techniques it will take to get from here to there. Imagine that you are Daniel Boone, leading a contentious band of bickering explorers westward. You climb a tall tree and can see the Cumberland Gap fifty miles away. All the explorers take off in different random, though generally westward, directions. Heaven only knows which path through the trees any one of them will take, but you still know that most of them will wind up going through the Gap.

We can see, in broad outline, a set of stages that we may go through, though not necessarily. There are many paths to the ultimate capabilities. However, this is one that seems reasonably likely, and includes Some well under Stood examples as landmarks.

Stage I

Essentially what we have now *nanoscale science* and technology including the ability to image at the atomic scale with scanning probe microscopes, and a very limited ability to manipulate, that is, by pushing things around with the same scanning probes. A scanning probe is essentially like feeling something with a stick. Because you have a computer behind it, you can touch it in a very close grid of points and produce a picture. Also at Stage I are capabilities like chemistry. That is to say, you can produce an atomically precise product, like a molecule, but only by mixing chemicals or similar bulk processes. Thus you can have molecules whose atoms are precisely arranged, but the molecules themselves are dumped in random piles.

Stage II

Now suppose we can take atomically precise parts, which we can form using chemistry or molecular biology. At the molecular scale there is a phenomenon called "*self-assembly*," which means that molecules will stick together in somewhat ordered ways. They act as if they were Lego blocks, but with little magnets instead of the pegs

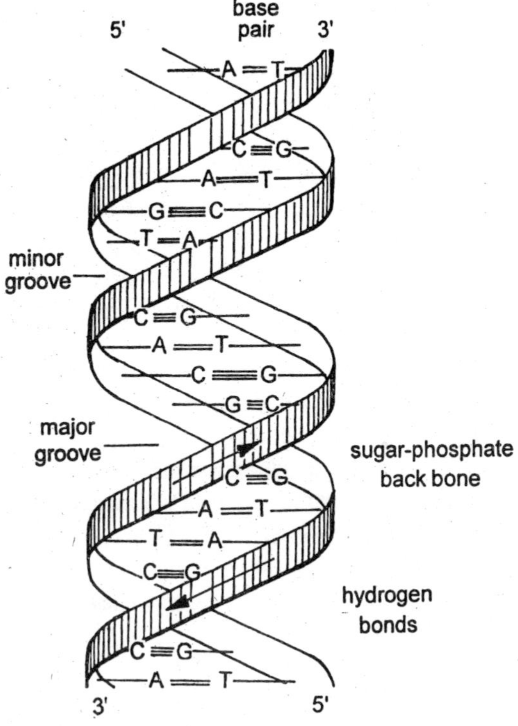

Fig. 1.3. DNA molecule.

and holes. You could take a bucket full of such blocks, shake it vigorously, and expect the blocks to fall together in some semblance of structure. It was originally envisioned that designed protein molecules might be made to do this kind of thing; it has been done with tailored DNA molecules and also smaller molecules of a general chemical nature. Stage II is still in the lab, but some experiments have shown the ability to make larger structures with planned, complex patterns.

Stage III

The next thing to do is to put the molecular Legos or Tinkertoys together, not just into pretty patterns but into functioning machines; in particular, machines that can put the Tinkertoys together themselves. This is the level of animal life: we are made of amino acid Tinkertoys, which our cellular machinery can put together to make more cellular machinery. At this stage we still require an externally created supply of the parts; we need proteins in our diet. Stage III will be the point where large—scale nanotech systems become feasible.

Stage IV

The next stage involves the system being able to make the Tinkertoys from simple molecules, as plants do. This is the stage where we expect to see the beginning of a long-term trend of falling costs. Before this, the inputs to the process are expensive products of sophisticated chemistry and *molecular biology*; afterward, they are common and cheap.

Stage V

The final stage is when the ability to make parts from simple molecules becomes general. In other words, instead of designing a part from the Tinkertoys, it's custom designed *atom by atom*. It's the difference between a machine made of actual Lego blocks and one made of custom-designed parts made with the same technology used to make like Lego blocks. This is considerably more challenging technically. It is also the stage where the more remarkable capabilities of nanotechnology, as compared to life, for example, will appear.

All the work in nanotechnology, and today's nano scale technologies, whatever its proximate goals, is leading in one general direction: toward the ability to build things smaller, with greater precision. Thus all die different paths ultimately lead to the same place: a technology where we can design things atom by atom, and build them as specified. What's more, we have an example of such a technology: it's biotechnology. The DNA blueprints and all the molecular machines inside the cell are built with atomic precision. So we know it's possible.

The science, as distinct from the engineering, of what happens at the atomic scale has been well established for half a century. It's the quantum mechanical wave equations of the same Erwin Schrodinger who wrote *What Is Life*. The equations tell you how *electrons* will arrange themselves to form atoms, and account for most chemical phenomena. In 1950, they were something of an ivory tower accomplishment because the techniques of the day couldn't solve them for most situations of interest. Nowadays, however, with modern computers, we can, at least enough to be extremely useful. So we can use this knowledge to design and simulate the machines that will be possible at the far confluence of the various paths that today's nano scale science and technology are taking.

Technological Revolutions

Normal science, for example, often suppresses fundamental novel ties because they are necessarily subversive of its basic commitments.

In these and other ways besides, normal science repeatedly goes astray. And when it does then begin the extraordinary investigations that lead the profession at last to a new set of commitments, a new basis for the practice of science. The extraordinary episodes in which that shift of professional commitments occurs are the ones known in this essay as scientific revolutions.

Humanity has been through several technology-driven transformations before. The first two, significant fractions of a million years ago, happened so long ago they have shaped our biological evolution. They were stone tools and the techniques for using fire. A third major technology; agriculture, has been around long enough for some evolutionary adaptations to start, but they are not complete. On the other hand, agriculture has had a profound influence on the social structure and physical circumstances of humanity and was a necessary precursor to urban civilization. The various technologies involved in making clothing had a similarly profound effect, allowing humans to live in places like Europe and North America.

Ships and wheeled carts made possible significant commerce. Reading and writing extended the spread of knowledge beyond the meager reach of the human memory. The scientific method itself was an invention, perhaps the most important one of the past millennium. The printing press takes first place among physical machines. Science and printing went hand in hand, enabling the industrial revolution and ushering in the modern world.

Nanotechnology has the potential for increasing our physical capabilities more than did the industrial revolution, expanding our ability to learn and communicate more than did the printing press, accelerating our ability to travel more than did the boat or the wheel, and enlarging the range of places we can live more than did clothing. It could induce greater biological changes in the human organism than the difference between humans and chimpanzees, indeed, greater than the difference between humans and horseshoe crabs. It is coming, possibly in the next decade, probably in the next twenty-five years, almost certainly in the twenty-first century.

2

Trivial Application of Nanotechnology

Nanotechnology changes just about everything—or does it? Substantial numbers of Amish and Mennonites living in the United States stick as close as they can do to a way of life frozen over a hundred years ago. Advancing technology has changed little for these folks—they still buy buggy whips! In the coming nanotech era, some individuals or groups of people might try to live a life based on current standards, using a minimal amount of the new technology. Advanced *nanotechnology* will give us the tools to alter the molecular make up of our own bodies, thereby generating a range of variously viable species, but if some of us choose to remain biologically human, it is fairly easy to imaging goods and services that we would want and how "*trivial*" applications of nanotechnology might provide them.

Trivial applications of complex new technologies can result in a substantial business. People who purchased the *first computers* would, at the time, have been a ghast at the concept of children using many millions of computer instructions per second (MIPS) to play dollar market! Last year Nintendo and Sega split a "*trivial*" five billion dollar market! The same is true of word *processor programs* running to today's average (tens of MIPS) PC. By the standards of two decades ago, a powerful computing machine is being used for the ancient and relatively simply task of written human communication. Those who first ran computers with such capability would have been absolutely appalled at the prospect of a machine being used in such a trivial way, spending almost all of its time waiting for a key stroke. A simple mechanical typewriter would suffice for the task.. Even a pencil would do. Using a million-instructions-per-second machine to type is technological overkill—but cost effective.

Gasoline Trees

If we consider future applications for nanotechnology, a similar overkill use would be a "*gasoline tree*," or, at first, an organic waste-to-gasoline converter. In the early stage, such a device might be the size and shape of a washing machine. You lift the lid and drop in food waste, paper, cardboard, chunks of wood, and so on. An ash hopper off to the side fills up with ceramic marbles, excess water runs down the drain, and carbon dioxide comes out a stack.

Organic materials placed in the machine would be rearranged by *nanomachines* forming liquid fuels suitable for the use in internal combustion engines. Actually, everything in this idea could be achieved today on an industrial scale; the chemistry and physics are sound, it just needs a little engineering development to put it into the home.

Seed

Later versions might come as a *seed*. Plant it next to the driveway, and it grows into a nice looking tree, complete with a recessed filler hose. Instead of continuing to grow more tree after it matures, it makes gasoline and stores it in the trunk. In fact, a vine that grows in the Amazon produces enough oil that oil can be skimmed off the tapped sap and used to run diesel engines. So, a gasoline tree might be possible to develop simply by using genetic engineering. And if you wonder how roads would be maintained without fuel taxes being paid, consider, a self-repairing road that used solar energy to grow more road in place—kind of like crabgrass. If the "road plant" were partly derived from kudu, road crews might be less busy with repairs and more concerned with keeping the road plant from overgrowing everything.

Energy

Providing *energy* is another trivial use of nanotechnology—or at least, providing the levels of energy currently used in the West. It is a good bet that solar collectors that are close to 100 percent efficient can be made with nanotechnology (well, ok, so they only hit 50 percent). Moreover, they could be self-repairing, grow from a seed like the gasoline tree, and double as permanent roofing. To install the solar collector, you would place a small square on your roof. A week or so later, it would grow an extension cord into your electric box and start supplying part of your electricity—the inverter is part of the distributed design. More elaborate models could supply a steady supply of electricity by growing an energy-storage module underground. Dense housing (if people still live that way) and large-scale industrial activities (whatever they are) will still require more energy than can be supplied by local sunlight, so good reasons remain for keeping the power grid operational.

Transportation

Regarding *transportation*, it is not entirely obvious that the conventional needs of a community for getting around—such as commuting to work—make sense in an area of nanotechnology, but if people want to congregate in large numbers (for rock concerts?), it would be a simple matter to provide a personalized subway stop for every house that wanted one. With a little notice, a new line could be run wherever and whenever a large group of people needed to be transported to a new location. A hotel/convention center could consist of a big tank of undifferentiated "*nanostuff*", a large area of unoccupied land, and a subway terminal. Choose one of thousands (millions) of designs for your convention.

Another possibility is *movable real estate*. With hordes of minute earth movers, surface structures could be moved slowly to open up new space for public works (again, whatever that might be in such an era). If an evolving community wanted to live in a less dense mode, all the structures, trees, roads, ultimates, and the like could creep outward at an inch or so a day.

A local version of the same idea would be timesharing part of your local real estate. A week before your big party, your house and lot would start to grow, while the neighbours' house would shrink and move away. By the time your party began, your house would have metamorphosed into whatever size mansion was needed. Over the next week, it would shrink to its original size—or perhaps a little smaller, if someone else were planing a party.

Home Appliances

Inside the home, there are countless applications for *trivial nanotechnology*. To start, the whole house could be carpeted with a self-cleaning rug. An early version would ripple like the cilia in your lungs to move all the little stuff that fell on it to a central bucket. Coins, nuts and bolts, paper clips, and the like could be sorted from dirt, dust, hair, skin flakes, and similar debris. A later version might be smart enough to know what it should eat. (It wouldn't do to have the rug eat the fur off one side of the cat while she took a nap!) A still later version might deal with the classic clothes-on-the-floor problem that has broken up so many marriages. Left for more than a few minutes, underwear, socks, even dress clothes could be disassembled by the carpet and rebuilt—clean, or course, and folded in dresser drawers, or hanging in the closest.

Clean floors are even more of a problem in the kitchen. Carpets for kitchens and bathrooms (and in all rooms for those with pets or babies) could also have soak-up powers. A network of pumped veins within the carpet could either be connected to the drain, or (depending on how you felt about it) attached to the food synthesizers. A nanotech carpet could be covered with cilia so short that it would look and feel like a no-wax surface. An active no-slip option could hang on to your shoes and keep them from slipping while offering no resistance to vertical movement. (The same technique could be used between tires and the roads to get greater-than-one-g performance from vehicles.)

Redecoration

Nanotechnology would also let you redecorate—daily. Walls and ceilings would get nanotech paint that could act as a universal display screen programmed to appear however you wanted, up to a very high definition TV screen. Talk about trivial, you could have stripped paint! Don't like the colour, or the art work? No problem: any colour, any design (delivered by optical fiber) can appear at your whim with a few clicks on whatever is being used as a computer interface. (Naturally, there will be a small charge for displaying copyrighted art works). You could even have a real-time view of the Grand Canyon on the wall of your living room.

Furniture could be ordered in the same way. Static items that you wanted to keep for a while could be built by your general-purpose household constructor. More dynamic items—an end table today and has sock tomorrow—would be made of general purpose nanostuff, sort of an "*Erector Set-in-a-bucket.*" And do not be too surprised when you come home to find that the kids have reprogrammed the spare bed into a slobbering, eight foot-high Godzilla.

Structurally, houses could be reinforced to the point that no natural disaster—short of a large incoming meteor—would do much damage. Certainly earthquakes, tornados, hurricanes, and the like could be reduced to minor annoyances. When buildings are reinforced with intelligent diamond fiber, such events would cause little or no real damage. And if damage. And if damage were done, a nanomachine-based house could repair itself.

If we choose to retain historical structures "as is," we can coat them with diamond, and all deterioration—termite damage included—would be stopped. Actually, every *prenanotech structure* might be considered historical, but most of them will be replaced, or greatly modified. If there is any question about the historical value of a structure, we can make a record of how it was put together (down to the marks on the nails). Then we can always reassemble it later.

With all the new comforts of home, will people leave home very often? This is a hard one to answer. Even now, a growing number of people choose to work at home and telecommute. In the future, if you "work" at home and if tasty, nutritious food is made in a "cabinet beast"—from elements scrounged off the floor or out of the air, and rearranged with energy from your roof top—and if must your shopping needs are met with home-shopping networks that deliver instruction sets by optical fiber to the *nanomachines* in your house, most of the reasons for going out vanish. Perhaps this is the real solution to the traffic problem!

Health Assurance

With all the time on our hands from less commuting, would we spend more of it worrying about our health? One of the primary reasons to develop nanotechnology is to cope with nasty health problems—including aging. Once we have done that, those who stick to a more-or-less standard human configuration will want improvements. Self-repairing, diamond-reinforced teeth for example. Or replacements for glasses. We are already reshaping eye lenses with crude surgery. Actively controlling lens focus would be an obvious application for *smart nanoimplants*. Floaters could be cleaned out of the eye fluid by nanomachines designed for the purpose.

A more complex modification would be rearranging the retinas of our eyes. Why evolution wound up with vertebrate nerve circuits in front of the light receptors is not well understood. We get along fine with the wires on the "*wrong*" side, but what a trivial use of vast developments in biological *nanoscience* to be able to turn the retina over in place. In that way we could get rid of "*unsightly*" blind spots. A more challenging problem is that of giving sight to people blind from birth. Building eyes may be easy compared to rewiring brains.

Given an understanding of biology at the *nanolevel*, *nanotechnology* should be up to solving the subset of human health problems due to mechanical or chemical causes. The same will be true for our pets. Given the rate at which new treatments are approved, our pets may get it first. The biggest difference for humans between horses and cars is that cars don't need attention every day. Members of the Society for Creative Anachronism (SCA) really need switchable animals. Their battles and pageants are short on horses, but few of them want to take care of a horse between events.

In addition, the people get so banged up in mock battles that the SPCA would object if real animals were subjected to the same beatings. Once a thorough understanding of human anatomy and chemistry is stored in *nanomachines*, the SCA can stage real—dripping with gore—*Conan-style* battles. After the performance, the chunks of the participants can be stuck back together, *Valhalla style*.

While there seems to be no limit to the level of *Medieval realism* that could be achieved, we don't have to put up with vast amounts of dung. A partial solution to the dung problem has been proposed in the form of a "doggy afterburner." Such a device would inhabit the lower intestine of an animal, burning the organics out of whatever came along. The nonorganic elements would be formed into ceramic marbles, to be excreted at rare intervals.

In a *nanotech world*, human relations with engineered animals could get very weird indeed. Today, pastoral nomads in Africa drink the blood of their cattle. A less messy method would be to grow plugs on the animals that could be connected to humans directly, supplying energy and materials to the bloodstream. Instead of killing sheep, you could bring in a batch and "*recharge*" from them. A lower-on-the-food-chain alternative could be a grafted on "*back-pack*" that would unfold when you lay in the sun, forming a large photosynthetic area. Assuming the nomad's sheep don't trample you, with thirty square meters of surface, a few hours a day soaking up rays would eliminate the need to eat animals or plants. Totally recycling "*hypervegetarians*" would be able to live the simple life par excellence—really leaving "nothing but footprints."

What Work?

In a world of *nanotechnological systems*, what would people do for work anyway? Work is an activity specific to humans that got started long, long ago when we started moving food from where we found it to where we could eat it. Work has become so elaborate, the origins have almost been lost and have been made very indirect by the invention of money.

One of the difficulties we face with nanotechnology is the question. "What do you do with the money, or its equivalent, that you receive in exchange for work?" There are still many things you might want to buy—land, transportation, elements, energy, information, be very expensive? There is a whole lot of land, especially with nanotech to make it worth something. It also seems likely that we could make a lot of really nice new land, reorganizing what we have here to make more beaches, or we could go into space. The reorganization can incorporate as many vacuum tunnels for transportation as we wanted. The useful elements are common and, with solar energy on tap, power for personal use can't be expensive. (Energy required to launch interstellar crafts may be another matter.) The kicker in a nanotech world is that no mater how much something costs to make in the first place, it costs almost nothing to duplicate, because it's mostly made of information. Finished goods will become "services" rather than distinct products. On the other hand, demand for personal services may actually increase, if you seek interaction with a real human.

We may see a long period of deflation offset, perhaps, by a reduction in the need for money to live. Or, at worst, you might get a photosynthesizing modification and live off the land. But what of money itself? Perhaps it will need to be based on chips that actively resist duplication, or possibly coins of genuinely rare elements. Or perhaps, continuing a growing trend, money will be no more than abstract patterns exchanged by computers.

Even if we don't need much money, what kind of work might be in demand? Is there a trend in the past few centuries (in this century?) that might continue? At the beginning of this century, well over half of American heads of families were farmers. Today some two percent are so occupied. Are the rest unemployed? Far from it. They work in a vast array of jobs, many of them information related, few of which existed at the turn of the century.

If you lump communications, entertainment, news and journalism, scientific endeavors, engineering, and most of the functions of government together and call it the "*information industry*," it is clear that this segment of the economy has accounted for an ever increasing fraction of what people do for employment. It may turn out that the nanotech revolution hardly causes a hiccough in this curve. If your "work" creates new knowledge, stories, or entertainment of any kind, you are likely to have a job in a nanotech world—if you want one. Government, too, faces uncertain prospects. My libertarian bent prefers little or no government, but my guess is that there will be a substantial fraction of the population working for the government, or perhaps living on welfare, provided with food, shelter, and endless reruns. Another trend in the service sector that gives me the shakes is more lawyers: the richer people get, the more legal actions they seem to be involved in.

Real Wealth

There is no reason a nanotech world should not be very rich. Again, this is in the line with long-term trends. Citizens of the United States who live in poverty today are (on average) better off in many ways (square feet of living area, nutrient content of food) than the average worker in the 1930s or those living under current average conditions in Eastern Europe. Food and floor space should be nearly trivial cost items in a nanotech world—at least for populations within an order of magnitude of our current population.

In an *era of nanotechnology*, it might turn out that people work because they want something to do, or are seeking status, rather than simply facing the harsh realities that say we must work or starve. More and more people have been deciding what they will do with their life by criteria other than just making enough money to survive. This may represent a long-term trend. Galactic dispersal of the nanotechnologically advantaged presents an interesting challenge that I will leave you with: How can we carry on economic exchange if the only thing exchanged is information carried on laser beams? Establishing a value for something where a bargaining cycle takes millennia is an intriguing problem. We may choose to send short-duration personality constructs designed to handle the bargaining along with encrypted goods. But would that be satisfying? And if so, for whom?

3

Evolution of Nanotechnology

The history of the development of the nanotechnology concept, from its roots in physics, chemistry, molecular biology, manufacturing and systems engineering, through its emergence at the Massachussetts Institute of Technology (MIT) will be dealt here. Theoretical and experimental work are surveyed, contrasting the top-down approach to miniaturization with the new bottom-up nanotechnology approach. The evolution of the molecular nanotechnology concept is stressed with respect to its role as a path to molecular-scale miniaturization and manufacturing.

Nanotechnology is defined as "the projected ability to use positional control of chemical reactions to build complex materials and devices (including molecular machinery) resulting in precise control of the structure of matter at the molecular level." Such materials and devices have been termed *eutactic*. Nanotechnology brings together results from many fields of science and engineering. The goals described are sometimes termed *molecular nanotechnology* to distinguish them from generic nano scale technologies.

Components of Evolutionary Concept

The evolutionary concept of nanotechnology entails an understanding of atoms, chemistry, atoms as a part of molecular machinery, and programmable manufacturing systems.

Atoms

Nanotechnology is based on a view of *atoms* as physical objects having definite mechanical properties and patterns of interconnection. The Greek philosopher Democritus (ca. 460-370 B.C.) proposed the

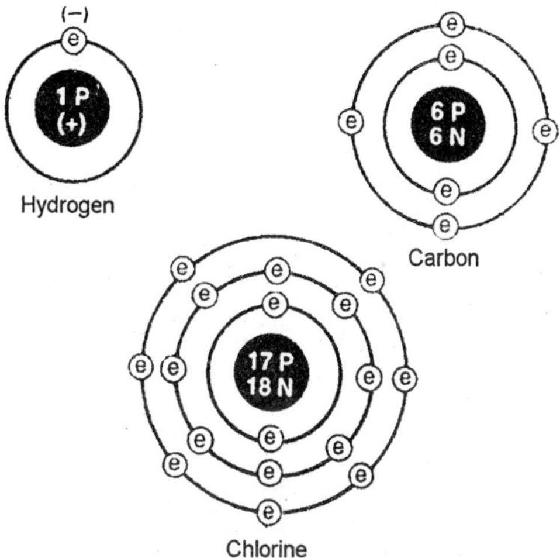

Fig. 3.1. Atomic structure.

idea of atoms: invisible, absolutely small (the name atom derives from *atomon*, "*indivisible*"), incompressible, and homogeneous. Atoms were seen as eternal. While their arrangements might change, the atoms themselves would be unaffected.

Lucretius, a Latin philosopher who lived in the first century B.C., developed *atomic theory* further in his work *De rerum natura* (*On the Nature of Things*). Infinite in number, but not in kind, atoms were said to differ only in shape, size, and weight. They were hard, eternal, and could not be divided into inseparable parts, though larger atoms had more such parts. Objects were described as "*systems of moving atoms.*"

While the ancients proposed atoms as small objects connected in varying combinations, they did not suggest that *small machines* might be made from these objects. Unlike the study of the natural world, the building of useful machines was regarded by the Greeks as a task for slaves. Residues of this attitude, favouring science over engineering, can be seen to this day and continue to affect the pace of technology development.

By 1803, John Dalton (a British chemist and physicist) had determined the relative *weights of atoms*. He went on to propose his atomic theory: All particles of different elements have different weights and sizes, and chemical combinations involve atoms connecting together.

The existence of atoms was denied by many scientists, even in the early twentieth century, including the Austrian physicist, Ernst

Mach, and the originator of physical chemistry, Wilhelm Ostwald. Their view (known as "*Positivism*") held that because atoms could not be seen, they must be a "*convenient fiction*" at most—a view that did not encourage the idea of building molecular machinery. Eventually, however, the weight of evidence overcame even these philosophically based objections and atoms soon were regarded as real objects.

Chemistry

Nanotechnology can be viewed as an elaboration of chemistry in which large molecules serve as machines that guide the synthesis of other large molecules. Chemistry and chemical synthesis have laid the foundations for this effort, but again, only after passing a series of conceptual hurdles.

In 1828, Friedrich Wohler achieved the *first synthesis* of an organic compound, *urea*, from inorganic materials. This overthrew the former assumption that organic molecules could be made only by organic (that is, biological) processes. Two important inferences—that carbon atoms make four bonds and that they can bond to each other—were made independently by Friedrich Kekule and Archibald Couper in 1858. In 1874, Jacobus Van't Hoff proposed that the four bonds of carbon were arranged in a *tetrahedral shape*. He showed that this explained how two compounds having the same chemical formula could have different bonding structures (such as right-handed versus left-handed stereoisomers). The idea that molecules have definite three-dimensional structures, was ridiculed at the time by the prominent chemist Adolph Kolbe as "*fanciful nonsense*" and "*supernatural explanations.*" By 1894, Emil Fischer had taken a next step by determining that enzymes could distinguish between stereoisomers, which led him to propose that substrates fit into enzymes in a *lock-and-key* relationship, With this

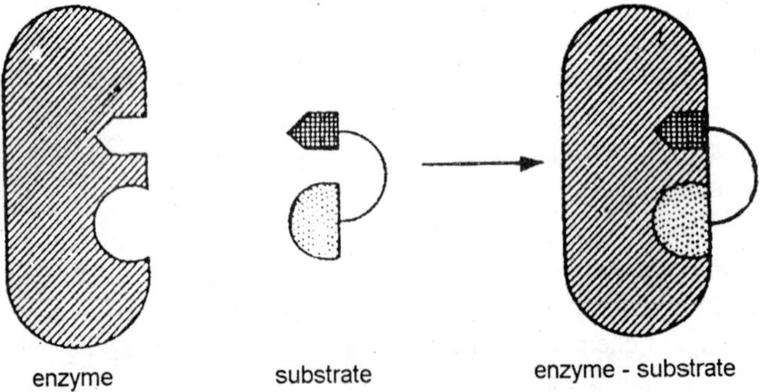

Fig. 3.2. Lock and key hypothesis.

determination, the idea of molecules as distinct, interacting objects with sizes, shapes, and mechanical properties was well-advanced.

In 1926, physicist Erwin Schrodinger put forward the foundations for a *quantum mechanical theory* of molecules and chemical bonding. He later wrote a book that explained life as based on molecular objects and molecular machinery.

Molecular machinery requires large molecules but the idea of *macromolecules* (molecules with molecular weights in the tens of thousands) had its encounter with skepticism during the 1920s and early 1930s. For example, leading chemists found it impossible to conceive that natural proteins could have molecular weights as large as the 16,000 claimed for *oxyhaemoglobin.* In response to the case for macromolecules, one organic chemist declared "that it was as if zoologists 'were told that somewhere in Africa an elephant was found who was 1,500 feet long and 300 feet high." In the 1980s, the concept of nanotechnology met with objections having a similar degree of scientific relevance.

Molecular Machinery

Molecular-scale machinery—that is, machines with every *atom bonded* in a precise manner, machines that hold work pieces in place to enable chemical reactions to occur—will be fundamental to nanotechnology. Examples are already seen in nature. An early discovery of this sort was the storage of information in the molecular tapes of DNA. Soon after followed a blossoming of the fields of molecular biology and biochemistry, with rapid progress in the understanding of natural molecular machinery throughout the 1950s, 1960s, 1970s, and beyond.

Throughout this period, an understanding of self-assembly developed—that is, individual molecules in solution can come together to form complex molecular machinery.

Programmable Manufacturing Systems

Another set of technological roots for the concept of nanotechnology is the development of *macroscopic machinery* for programmable manufacturing systems. First introduced in 1805, looms operating from instructions on punched cards were perhaps the earliest programmable manufacturing machines. Charles Babbage, working from the 1820s through 1871, designed mechanical calculating machines based on moving parts made of brass. He started to build them, but ran out of funds. Recently, a Babbage design was constructed and found to work, even given the limitations in precision that he faced at that time. The

Table 3.1. A comparison of macroscale and biomolecular components and functions.

Device	*Function*	*Molecular example(s)*
Struts, beams, casings	Transmit force, hold positions	Microtubules, cellulose
Cables	Transmit tension	Collagen
Fasteners, glue	Connect parts	Intermolecular forces
Solenoids, actuators	Move things	Conformation-changing proteins, action/myosin
Motors	Turn shafts	Flagellar motor
Drive shafts	Transmit torque	Bacterial flagella
Bearings	Support moving parts	Sigma bonds
Containers	Hold fluids	Vesicles
Pumps	Move fluids	Flagella, membrane proteins
Conveyor belts	Move components	RNA moved by fixed ribosome (partial analog)
Clamps	Hold work pieces	Enzymatic binding sites
Tools	Modify work pieces	Metallic complexes, functional groups
Production lines	Construct devices	Enzyme systems, ribosomes
Numerical control systems	Store and read programs	Genetic system

original project's failure is blamed partly on "the cultural divide between pure and applied science." Since the idea of *mechanical computation* played a key role in the development of the nanotechnology concept, Babbage's failure may have delayed not just computation, but also nanotechnology.

Computer technology, combined with power-driven machinery has made possible today's numerically controlled machine tools. These flexible manufacturing systems at the macro scale, in turn, helped inspire the idea of nano scale molecular assemblers.

Partial Synthesis by Feynman

During an after-dinner talk in 1959, physicist Richard Feynman discussed a path of miniaturization in which large machines could be used to make smaller machines, which would then build smaller machines, and so on, toward the molecular level. *Microscopic lathes* were envisioned, along with problems in making microscopic automobiles for mites. This would be a "*top-down approach*" (similar to today's semiconductor technology) in that standard materials would be sculpted into a desired shape. The resulting tools, while small, would not be made with molecular-scale control; they would not be eutactic.

Toward the end of his talk, *Feynman* hinted at a "*bottom-up approach*," in which control would extend to the molecular scale. "The principals of physics, do not speak against the possibility of maneuvering things atom-by-atom," he said. "But, it is interesting that it would be, in principle, possible for a physicist to synthesize any chemical substance that the chemist writes down. Give the orders and the physicist synthesizes it. How? Put the atoms down where the chemist says and so you make the substance."

Feynman's visionary talk inspired a generation of engineers, those working in *semiconductor miniaturization* and those building micro machines using similar technology. However, this vision of non eutactic machines building chemicals did not extend to molecular nanotechnology. Almost two decades would pass before the proposal was made that manufacturing machinery of molecular precision could be used to build other materials and devices of molecular precision. That proposal, when it came, arrived from a different direction.

Methodology Required

Feynman's talk was an example of a process familiar to the *aerospace community*—that is, taking known scientific principles and projecting technological advances that can be made within the bounds of those principles. This process has been termed *theoretical applied science*. Examples from the aerospace field include Goddard's rocket designs, Clarke's geostationary satellite proposal, and the early Apollo designs.

Theoretical Applied Science

A theoretical applied science approach was necessary in order to join the components described previously—atoms as objects, chemistry, molecular machinery, and programmable manufacturing systems—to form the concept of molecular nanotechnology. Furthermore, two particular emphases were needed (both of which are also familiar in the aerospace field): a decades-long planning horizon and a focus on *systems engineering*. Given these methodological similarities, it is perhaps not surprising that the nanotechnology concept arose from within the aerospace community.

Nanotechnology Concept Emerges at MIT

As a student in MIT's Department of Interdisciplinary Science in the late 1970s, K. Eric Drexler's studies centered on the design of manufacturing systems for use in the novel environment of space. This work applied a systems-engineering approach to manufacturing systems, including chemical systems, designed to operate in novel environments, combined with a research focus aimed at under standing long-term technological capabilities. Topics studied included computation, manufacturing, materials science, biology, and design engineering.

A focus on the nanometer scale was provided by Drexler's undergraduate thesis, on lightsails using reflective sheets of aluminum as thin as 20 to 100 nm. A side interest in the exploding field of molecular biology provided information on the molecular machines found in nature. In late 1976, some of the concepts required for nanotechnology came together in the form of what would later be termed *biocomputers* (computers based on self-assembled structures of biomolecules). These ideas included, for example, *lipid bilayers* used as insulators.

In the spring of 1977, consideration of self-assembled systems of molecular machines based on biological models led to Drexler's realization that such machines could be used to position reactive molecules—guiding chemical reactions under programmable control to build complex structures, not unlike the operation of a ribosome. This was the core concept of molecular nanotechnology: the design of manufacturing systems for use in the novel environment of the molecular scale world.

Drexler's studies were continued informally in parallel with his pursuit of a graduate degree in space industrialization supported by a National Science Foundation graduate fellowship. This combination proved incompatible, so work was continued during a research affiliate appointment, first at the MIT Space Systems Laboratory and later at the MIT Artificial Intelligence Early work on scanning tunneling microscopes immediately made clear that these instruments might be useful in developing nanotechnology In 1982, Drexler made an early effort to explain the concept to a lay readership.

Distinctions between the *molecular machinery* of nanotechnology and *molecular electronics* were drawn in a paper presented by Drexler in 1983 at the "Second International Symposium on Molecular Electronic Devices." Drexler described *aerospace applications* in 1983 at a presentation during the second "International Space Development Conference" held in Houston, TX. Medical applications were also explored by Drexler during this time, but the medical community was not ready for the concept. A paper by Drexler invited by an editor at the *Journal of the American Medical Association* was dismissed by a referee as "*science fiction.*" The dismissal of engineering concepts in similar terms, like wise without supporting scientific criticism, is familiar from the history of the spaceflight movement.

A lecture at MIT by Drexler in January 1985 led to the formation of the MIT Nanotechnology Study Group by a group including Drexler, myself, and Christopher Fry (then with the MIT Artificial Intelligence Laboratory). Fry led the study group for several years, followed by David Forrest (then a doctoral candidate with MIT's Department of Materials Science) and later by David Lindbergh (a software developer).

THEORETICAL WORK

By June 1986, when the first book on nanotechnology was published, the center of interest in nanotechnology had moved to the Silicon Valley (California) area. The Foresight Institute, a nonprofit foundation with the purpose of studying and guiding nanotechnology's societal impact, was founded in 1986 in Palo Alto, CA, by Drexler, myself, and James C. Bennett (then vice president of the American Rocket Company, a private launch firm).

An early design for a *mechanical nanocomputer* was presented in 1986 at the "Third International Symposium on Molecular Electronic Devices." The basic mechanism of the design was termed *rod logic*. The analysis included static and sliding friction, thermal noise, overall energy dissipation, and chemical stability. In November 1987, an analysis of early designs for *nanomechanical gears* and *bearings* was presented at the "IEEE Micro Robots and Teleoperators Workshop." The nanotechnology concept started gaining recognition when it first appeared in Scientific American in January 1988.

In the spring of 1988, Drexler taught the first university class in nanotechnology to about 50 students under the sponsorship of Stanford University's Department of Computer Science, where he was a visiting scholar. Potential applications to supercomputing were described at the "Third International Conference on Supercomputing in May 1988."

Designs of molecular components continued in the summer of 1990 with the invention by Ted Kaehler (then a visiting researcher with the Foresight Institute on sabbatical from Apple Computer) of molecular-bracket-based approach to the design of irregular structures. A fundamental problem in molecular engineering is the design of structures able to hold an object at any desired point in space relative to another object, with good accuracy. The atomic graininess of matter precludes simply cutting a bracket of the correct size and shape, and designing a bracket of the right shape when needed could require a considerable search among options. The Kaehler bracket approach begins by generating a large library of bracket-structure designs to provide a wide range of lengths, angles, and so forth. These designs are then indexed according to their geometrical properties, so that a structure of the right size and shape can be selected whenever needed. Development of a Kaehler bracket library solves a broad set of problems demanding the construction of irregular shapes constrained by tight tolerances.

In 1990, Ralph Merkle began work in computational nanotechnology at the Xerox Palo Alto Research Center. This work has included the analysis of nanomechanical and electronic computational systems that can approach *thermodynamic reversibility*. Merkle also has collaborated with researchers at the California Institute of Technology to analyze reactive tools useful in the molecular manufacture of diamond structure using the techniques of quantum chemistry. He has also worked in the design of various nano mechanical components using molecular modeling software, some performed in collaboration with Drexler and some with Lakshmikantan Balasubramaniam of MIT. These designs include several molecular bearings and a planetary gear. In 1992, Merkle produced a video animation showing the dynamics of these systems. In a 1992 paper, IBM researcher Jeffrey Soreff explored a method for construction of diamond structures using a relatively *mature nanotechnology*.

The first technical book appeared in 1992 and covered relevant physical laws and their application to nanoscale systems, design rules, and experimental approaches. An early version of this volume served as a dissertation for the first Ph.D. in the field of molecular nanotechnology. It was granted to Drexler in 1991 as part of an interdisciplinary program at MIT. The departments represented included Electrical Engineering and Computer Science, Chemistry, Biology, and Mechanical Engineering. The program was hosted by the Media Lab through a committee chaired by Marvin Minsky.

Further Work

The interdisciplinary nature of the development of nanotechnology makes a broad range of experimental work relevant. A comprehensive survey of such work would be too lengthy to present here. Therefore, a brief summary of a few prominent events follows.

Work in the United States

The construction of large designed molecules is an enabling technology for nanotechnology. In 1987, the first designed protein was produced by researchers at Du Pont. Since then, work on engineered proteins has continued to progress; an approach involving nonstandard amino acids has been described.

Proximal probe instruments can be used as positioning tools. In 1989, the scanning tunneling microscope was used at IBM's Almaden Research Center to position 35 atoms to spell out the company's three-letter logo, proving that atoms can be positioned with atomic precision. A similar result was soon produced in Japan by Hitachi. J. A. Armstrong (IBM chief scientist and vice president for science and technology) has said, "*Nanoscience* and *nanotechnology* will be central to the next epoch of the Information Age, and will be as revolutionary as science and technology at the micron scale have been since the early 70s. Indeed, we will have the ability to make electronic and mechanical devices atom-by-atom when that is appropriate to the job at hand."

At the U.S. National Institute of Standards and Technology, Clayton Teague is currently improving the precision of the STM's positioning abilities. A related instrument, the *atomic force microscope* (AFM), can be used for positioning and imaging nonconductive substances. This may prove to be a key enabling technology for nanotechnology.

Work on enabling science and technology was encouraged by the first research conference on the topic (the "First Foresight Conference on Nanotechnology") in 1989. It was sponsored by the Foresight Institute and Global Business Network, and hosted by Stanford University's Department of Computer Science. Researchers from many enabling fields reported on their work, including the first presentation of a proposal for a molecular tip array approach using the AFM.

The Institute for *Molecular Manufacturing* (IMM) was founded in 1991 in Palo Alto, CA. IMM is a nonprofit foundation with the goal of promoting nanotechnology research and is now making seed grants to researchers. IMM is projected to grow into a full-fledged research institute.

At the second "Foresight Conference on Molecular Nanotechnology" in 1991 (sponsored by the Foresight Institute, the Institute for Molecular Manufacturing, Stanford University's Department of Materials Science, and University of Tokyo's Research Center for Advanced Science and Technology, among others), presentations demonstrated that major progress had been made in the two years since the first meeting. Workshop sessions emphasized approaches involving a combination of proximal probes and designed molecules.

Work in Switzerland and Sweden

The *scanning tunneling microscope* was originally developed at an IBM laboratory in Switzerland. Researchers there have continued at the forefront of work in proximal probes. The "*Micronics*" program in Sweden, though small, has several projects of relevance to nanotechnology.

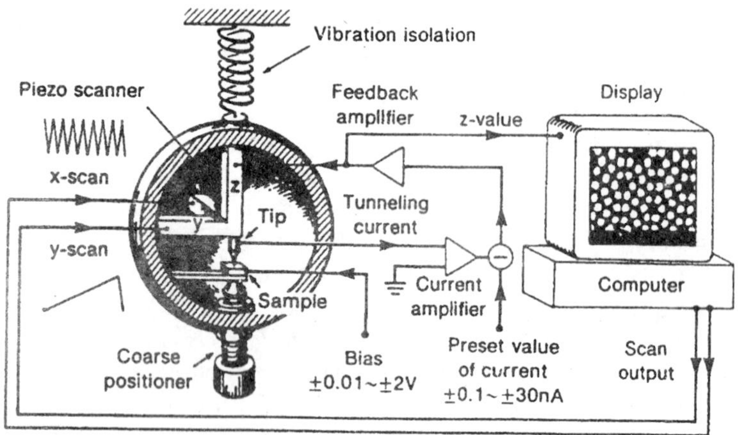

Fig. 3.3. Scanning tunneling microscope.

Work in Japan

Research in Japan often emphasizes project-oriented technology development, a useful methodology for pursuing nanotechnology. An interdisciplinary team of researchers is brought together (often from industry) for a number of years to pursue broad technological goals in a particular direction. At the Protein Engineering Research Institute (PERI) in 1990, this teamwork resulted in the construction of the largest (at that time) designed protein.

These organizational methods are used in the prestigious ERATO projects, funded by the Exploratory Research in Advanced Technology program within Japan's Science and Technology Agency. Many of these have had a molecular or nanoscale focus. Perhaps the most relevant has been the *nano Atom-craft Project*, which is attempting to use the scanning tunneling microscope for atom-by-atom construction of electronic devices.

A more focused, ambitious approach to nanotechnology in Japan has developed since a 1990 Ministry of International Trade and Industry (MITI) Symposium on the topic, which was organized around a visit by Drexler. In 1991, a planned investment of $185 million was announced by MTTI, to be spent over a period of 10 years on a nanotechnology project (in the broad sense of the word). In addition, the Agency of Industrial Science and Technology, also part of MITT, is establishing a new research center with a nanotechnology effort as a prime focus. In 1992, the British-based journal *Nature* held its first "Nano technology" conference in Japan.

Public Education and Policy Implications

In parallel with technical developments, efforts have been made to inform the public and policymakers regarding potential applications of nanotechnology and the effects of these applications. The Foresight Institute publishes Foresight Update, as well as conducting general conferences and research conferences on alternate years. The institute also responds to media inquiries, providing information for newspapers, magazines and television. A documentary on nanotechnology entitled "Little by Little," produced for Channel 4 in England, won the prestigious Prix du Centre National de la Recherche in 1990. A three-part series entitled "Nanospace" produced by NHK (Japan's public television organization) was aired in 1992.

In the late 1980s, J. Storrs Hall organized an online discussion group, named *sci.nanotech*, on the *computer network USENET*. The 1990 *Yearbook of Science and the Future* published by Encyclopedia Britannica included an article on nanotechnology. Also in 1990, Ted Kaehler formed a discussion group on nanotechnology within the Computer Professionals for Social Responsibility. In 1991, a second book on nanotechnology, *Unbounding the Future*: the Nanotechnology Revolution, was written for a general readership. A Japanese edition of *Engines of Creation* was published in 1992, translated by professor Masuo Aizawa of the Tokyo Institute of Technology.

Interest in policy examinations is increasing, in parallel with the state of information within the scientific community. World leaders in government and business were briefed in 1990 at the "World Economic Forum" in Davos, Switzerland. A foundation chartered to investigate nanotechnology policy options (the Center for Constitutional Issues in Technology) was formed in 1991 in Palo Alto, CA, headed by James C. Bennett.

Communication regarding these ideas has been hampered by confusions in terminology. Terms used in theoretical applied science to describe long-term goals are often borrowed by laboratory scientists to describe short-term goals and recent results. For example, in the U.S. the term "*nanotechnology*" has been used by those researchers doing *submicron semiconductor work*. This makes it difficult to discuss the goal and forces researchers to use the longer terms "*molecular nanotechnology*" and "*molecular manufacturing*." If history repeats, these terms will be borrowed and redefined as soon as they are recognized as describing attractive goals. In Britain, the term "*nanotechnology*" has been applied to a wide range of *submicron technologies* and *measurement techniques*. This paper cites an earlier paper referring to '*Nano-Technology*'—with single quotes and capital letters—as a prior usage of the term "*nanotechnology*." The earlier paper does not discuss anything resembling molecular nanotechnology.

Interest within the Space Development Community

Applications of nanotechnology to space industry have continued to be explored at space conferences such as the American Astronautical Society meetings and the "International Space Development Conferences (ISDC)." At the 1991 ISDC, a Molecular Manufacturing Shortcut Group was founded within the National Space Society to pursue these applications. Also at the 1991 ISDC, Drexler received the Space Pioneer Award in the Scientist/Engineer category, in recognition of his work in nanotechnology for space development.

Prospects for Nanotechnology

Several factors delay the development of nanotechnology. Barriers between disciplines slow progress. In some countries, applied science is regarded as less prestigious than basic science, and governments encourage the best technical minds to enter "pure" science rather than applied science. Accordingly, tool development is under funded and many scientists completely avoid tool development.

These difficulties do little to weaken the strong pressures pushing research toward nanotechnology. Skepticism regarding the nanotechnology concept is dissipating, as did earlier skepticism regarding the reality of atoms, molecules with distinct shapes and macromolecules, and the practicality of engineering proposals such as the airplane and rocket. Work in many fields—chemistry, biochemistry, molecular biology, physics, applied physics, materials science, computer science, mechanical engineering—is converging on thorough control of matter at the molecular level. The development of nanotechnology appears inevitable and will bring great improvements in high technology. The timing of these results will depend on the degree of well-focused effort that is brought to bear.

4

Opportunities and Risks of Nanotechnologies

The term *nanotechnology* describes a range of technologies performed on a nanometer scale with widespread applications as an enabling technology in various industries. Nanotechnology encompasses the production and application of physical, chemical, and biological systems at scales ranging from individual atoms or molecules to around 100 nanometers, as well as the integration of the resulting nanostructures into larger systems. The area of the dot of this "i" alone can encompass 1 million nanoparticles.

What is different about materials on a nanoscale compared to the same materials in larger form is that, because of their relatively larger surface-area-to-mass ratio, they can become more chemically reactive and change their strength or other properties. Moreover, below 50 nm, the *laws of classical physics* give way to quantum effects, provoking different optical, electrical and magnetic behaviours.

Nanoscale materials have been used for decades in applications ranging from window glass and sunglasses to car bumpers and paints. Now, however, the convergence of scientific disciplines (chemistry, biology, electronics, physics, engineering etc.) is leading to a multiplication of applications in materials manufacturing, computer chips, medical diagnosis and health care, energy, biotechnology, space exploration, security and so on. Hence, nanotechnology is expected to have a significant impact on our economy and society within the next 10 to 15 years, growing in importance over the longer term as further scientific and technology breakthroughs are achieved.

It is this convergence of science on the one hand and growing diversity of applications on the other that is driving the potential of nanotechnologies. Indeed, their biggest impacts may arise from unexpected combinations of previously separate aspects, just as the internet and its myriad applications came about through the convergence of telephony and computing.

Sales of *emerging nanotechnology* products have been estimated by private research to rise from less than 0.1 % of global manufacturing output today to 15 % in 2014. These figures refer however to products "*incorporating nanotechnology*" or "manufactured using nanotechnology". In many cases nanotechnology might only be a minor - but sometimes decisive — contribution to the final product.

Environment, Health and Nanoparticles

Along with the discussion of their enormous technological and economic potential, a debate about new and specific risks related to nanotechnologies has started.

The catch-all term "*nanotechnology*" is so broad as to be ineffective as a guide to tackling issues of risk management, risk governance and insurance. A more differentiated approach is needed regarding all the relevant risk management aspects.

With respect to health, environmental and safety risks, almost all concerns that have been raised are related to free, rather than fixed manufactured nanoparticles. The risk and safety discussion related to free nanoparticles will be relevant only for a certain portion of the widespread applications of nanotechnologies.

Epidemiological studies on ambient fine and ultrafine particles incidentally produced in industrial processes and from traffic show a correlation between ambient air concentration and mortality rates. The health effects of ultrafine particles on respiratory and cardiovascular endpoints highlight the need for research also on manufactured nanoparticles that are intentionally produced.

In initial studies, manufactured nanoparticles have shown toxic properties. They can enter the human body in various ways, reach vital organs via the blood stream, and possibly damage tissue. Due to their small size, the properties of nanoparticles not only differ from bulk material of the same composition but also show different interaction patterns with the human body.

At present, the exposure of the general population to nanoparticles originating from dedicated industrial processes is marginal in relation to those produced and released unintentionally e.g. via combustion processes. The exposure to manufactured nanoparticles is mainly concentrated on workers in nanotechnology research and nanotechnology companies. Over the next few years, more and more consumers will be exposed to manufactured nanoparticles. Labelling requirements for nanoparticles do not exist. It is inevitable that in future manufactured nanoparticles will be released gradually and accidentally into the environment. Studies on biopersistence, bioaccumulation and ecotoxicity have only just started.

What is Nanotechnology and What Makes it Different?

A nanometer (nm) is one thousand millionth of a meter. A single human hair is about 80,000 nm wide, a red blood cell is approximately 7,000 nm wide, a DNA molecule 2 to 2.5 nm, and a water molecule almost 0.3 nm. The term "*nanotechnology*" was created by Norio Taniguchi of Tokyo University in 1974 to describe the precision manufacture of materials with nanometer tolerances, but its origins date back to Richard Feynman's 1959 talk "There's Plenty of Room at the Bottom" in which he proposed the direct manipulation of individual atoms as a more powerful form of synthetic chemistry. Eric Drexler of MIT expanded Taniguchi's definition and popularized nanotechnology in his 1986 book "Engines of Creation: The Coming Era of Nanotechnology". On a nanoscale, i.e. from around 100 nm down to the size of atoms (approximately 0.2 nm) the properties of materials can be very different from those on a larger scale. Nanoscience is the study of phenomena and manipulation of materials at atomic, molecular and macromolecular scales, in order to understand and exploit properties that differ significantly from those on a larger scale. Nanotechnologies are the design, characterization, production and application of structures, devices and systems by controlling shape and size on a nanometer scale.

Modern industrial nanotechnology had its origins in the 1930s, in processes used to create silver coatings for photographic film; and chemists have been making polymers, which are large molecules made up of nanoscale subunits, for many decades. However, the earliest known use of nanoparticles is in the ninth century during the Abbasid dynasty. Arab potters used nanoparticles in their glazes so that objects would change colour depending on the viewing angle (the so-called *polychrome lustre*). Today's nanotechnology, i.e. the planned manipulation of materials and properties on a nanoscale, exploits the interaction of three technological streams:

1. New and improved control of the size and manipulation of nanoscale building blocks;
2. New and improved characterization of materials on a nanoscale (e.g., spatial resolution, chemical sensitivity);
3. New and improved understanding of the relationships between nanostructure and properties and how these can be engineered.

The properties of materials can be different on a nanoscale for two main reasons. First, nanomaterials have, relatively, a larger surface area than the same mass of material produced in a larger form. This can make materials more chemically reactive (in some cases materials that are inert in their larger form are reactive when produced in their nanoscale form), and affect their strength or electrical properties. Second, below 50 nm, the laws of classical physics give way to quantum effects, provoking optical, electrical and magnetic behaviours different from those of the same material at a larger scale. These effects can give materials very useful physical properties such as exceptional electrical conduction or resistance, or a high capacity for storing or transferring heat, and can even modify biological properties, with silver for example becoming a bactericide on a nanoscale. These properties, however, can be very difficult to control. For example, if nanoparticles touch each other, they can fuse, losing both their shape and those special properties—such as the magnetism—that scientists hope to exploit for a new generation of microelectronic devices and sensors.

On a nanoscale, chemistry, biology, electronics, physics, materials science, and engineering start to converge and the distinctions as to which property a particular discipline measures no longer apply. All these disciplines contribute to understanding and exploiting the possibilities offered by nanotechnology, but if the basic science is converging, the potential applications are infinitely varied, encompassing everything from tennis rackets to medicines to entirely new energy systems. This twin dynamic of convergent science and multiplying applications means that nanotechnology's biggest impacts may arise from unexpected combinations of previously separate aspects, just as the internet came about through the convergence of telephony and computing.

Nanomaterials: Basic Building Blocks

This section outlines the properties of three of the most talked-about nanotechnologies: carbon nanotubes, nanoparticles, and quantum dots.

Carbon nanotubes

Carbon nanotubes, long thin cylinders of atomic layers of graphite, may be the most significant new material since plastics and are the most significant of today's nanomaterials. They come in a range of different structures, allowing a wide variety of properties. They are generally classified as single-walled (SWNT), consisting of a single cylindrical wall, or multi-walled nanotubes (MWNT), which have cylinders within the cylinders. When the press mentions the amazing properties of nanotubes, it is generally SWNT they are referring to.

However, SWNT are more difficult to make than MWNT, and confusion arises about the quantities of nanotubes actually being manufactured. Carbon Nanotechnologies of Houston, one of the world's leading producers, only makes up to 500 g per day. One problem is that economies of scale are practically impossible with today's production technologies—the machines used to manufacture the tubes cannot be scaled up, so producing bigger quantities means using more machines.

Another drawback is that it is difficult to make *nanotubes* interact with other materials. For example, to fully exploit their strength in composite materials, nanotubes need to be "attached" to a polymer. They are chemically modified to facilitate this (a process known as "*functionalization*"), but this process reduces the very properties the nanotubes are being used for. In the long-term, the ideal solution would be to use pure nanomaterials, e.g. nanotubes spun into fibers of any desired length, but such a development is unlikely in the next couple of decades unless a radically more efficient production process is developed.

The most promising applications of nanotubes may be in *electronics* and *optoelectronics*. Today, the electronics industry is producing the vital components known as *MOSFETs* (metal oxide semiconductor field-effect transistors) with critical dimensions of just under 100 nm, with half that size projected by 2009 and 22 nm by 2016. However, the industry will then encounter technological barriers and fundamental physical limitations to size reduction. At the same time, there are strong financial incentives to continue the process of scaling, which has been central in the effort to increase the performance of computing systems in the past. A new microchip manufacturing plant costs around $1.5 billion, so extending the technology's life beyond 2010 is important. One approach to overcoming the impending barriers while preserving most of the existing technology, is to use new materials.

With *carbon nanotubes*, it is possible to get higher performance without having to use ultra thin silicon dioxide gate insulating films. In addition, *semiconducting SWNTs*, unlike silicon, directly absorb and emit light, thus possibly enabling a future *optoelectronics technology*. The SWNT devices would still pose manufacturing problems due to quantum effects at the nanoscale, so the most likely advantage in the foreseeable future is that carbon nanotubes will allow a simpler fabrication of devices with superior performance at about the same length as their scaled silicon counterparts.

Other proposed uses for nanotubes:

Chemical and genetic probes

A nanotube-tipped atomic force microscope can trace a strand of DNA and identify chemical markers that reveal which of several possible variants of a gene is present in the strand. This is the only method yet invented for imaging the chemistry of a surface, but it is not yet used widely. So far it has been used only on relatively short pieces of DNA.

Mechanical memory (nonvolatile RAM)

A screen of nanotubes laid on support blocks has been tested as a binary memory device, with voltages forcing some tubes to contact (the "on" state) and others to separate ("off"). The switching speed of the device was not measured, but the speed limit for a mechanical memory is probably around one megahertz, which is much slower than conventional memory chips.

Field emission based devices

Carbon Nanotubes have been demonstrated to be efficient field emitters and are currently being incorporated in several applications including flat-panel display for television sets or computers or any devices requiring an electron producing cathode such as X-ray sources (e.g. for medical applications).

Nanotweezers

Two nanotubes, attached to electrodes on a glass rod, can be opened and closed by changing voltage. Such tweezers have been used to pick up and move objects that are 500 nm in size. Although the tweezers can pick up objects that are large compared with their width, nanotubes are so sticky that most objects can't be released. And there are simpler ways to move such tiny objects.

Supersensitive sensors

Semiconducting nanotubes change their electrical resistance dramatically when exposed to alkalis, halogens and other gases at room temperature, raising hopes for better chemical sensors. The sensitivity of these devices is 1,000 times that of standard solid state devices.

Hydrogen and ion storage

Nanotubes might store hydrogen in their hollow centers and release it gradually in efficient and inexpensive fuel cells. They can also hold lithium ions, which could lead to longer-lived batteries. So far the best reports indicate 6.5 percent hydrogen uptake, which is not quite dense enough to make fuel cells economical. The work with lithium ions is still preliminary.

Sharper scanning microscope

Attached to the tip of a scanning probe microscope, nanotubes can boost the instruments' lateral resolution by a factor of 10 or more, allowing clearer views of proteins and other large molecules. Although commercially available, each tip is still made individually. The nanotube tips don't improve vertical resolution, but they do allow imaging deep pits in nanostructures that were previously hidden.

Superstrong materials

Embedded into a composite, nanotubes have enormous resilience and tensile strength and could be used to make materials with better safety features, such as cars with panels that absorb significantly more of the force of a collision than traditional materials, or girders that bend rather than rupture in an earthquake. Nanotubes still cost 10 to 1,000 times more than the carbon fibers currently used in composites. And nanotubes are so smooth that they slip out of the matrix, allowing it to fracture easily.

There are still many technical obstacles to overcome before carbon nanotubes can be used on an industrial scale, but their enormous potential in a wide variety of applications has made them the "*star*" of the nanoworld and encouraged many companies to commit the resources needed to ensure that the problems will be solved. Fujitsu, for example, expects to use carbon nanotubes in 45 nm chips by 2010 and in 32 nm devices in 2013.

Nanoparticles

Nanoparticles have been used since antiquity by ceramists in China and the West, while 1.5 million tons of carbon black, the most abundant nanoparticulate material, are produced every year. But, as mentioned earlier, nanotechnology is defined as knowingly exploiting the nanoscale nature of materials, thereby excluding these examples. Although metal oxide ceramic, metal, and silicate nanoparticles constitute the most common of the new generation of nanoparticles, there are others too. A substance called chitosan for example, used in hair conditioners and skin creams, has been made in nanoparticle form to improve absorption.

Moving to nanoscale changes the physical properties of particles, notably by increasing the ratio of surface area to volume, and the emergence of quantum effects. High surface area is a critical factor in the performance of catalysis and structures such as electrodes, allowing improvement in performance of such technologies as fuel cells and batteries. The large surface area also results in useful interactions between the materials in nanocomposites, leading to special properties such as increased strength and/or increased chemical/heat resistance. The fact that nanoparticles have dimensions below the critical wavelength of light renders them transparent, an effect exploited in packaging, cosmetics and coatings.

Quantum dots

Just as carbon nanotubes are often described as the new plastics, so *quantum dots* are defined as the ball bearings of the nano-age. Quantum dots arc like "*artificial atoms*". They are 1 nm structures made of materials such as silicon, capable of confining a single electron, or a few thousand, whose energy states can be controlled by applying a given voltage. In theory, this could be used to fulfil the alchemist's dream of changing the chemical nature of a material, making lead emulate gold, for example.

One more likely set of possible applications exploits the fact that quantum dots can be made to emit light at different wavelengths, with the smaller the dot the bluer the light. The dots emit over a narrow spectrum making them well suited to imaging, particularly for biological samples. Currently, biological molecules are imaged using naturally fluorescent molecules, such as organic dyes, with a different dye attached to each kind of molecule in a sample. But the dyes emit light over a broad range of wavelengths, which means that their spectra overlap and only about three different dyes emit light over a broad range of wavelengths, which means that their spectra overlap and only about three different dyes can be used at the same time. With quantum dots, full-colour imaging is possible because large numbers of dots of different sizes can be excited by a light source with a single wavelength.

The wide range of colours that can be produced by quantum dots also means they have great potential in security. They could, for example, be hidden in bank notes or credit cards, producing a unique visible image when exposed to ultraviolet light.

It is possible to make *light-emitting diodes* (*LEDs*) from quantum dots which could produce white light e.g. for buildings or cars. By controlling the amount of blue in the emission-control the "*flavor*" or "*tone*" of the white light can be tuned.

Quantum dots are also possible materials for making *ultrafast*, all-optical switches and logic gates that work faster than 15 terabits a second. For comparison, the Ethernet generally can handle only 10 megabits per second. Other possible applications are all-optical *demultiplexers* (for separating various multiplexed signals in an optical fiber), all-optical computing, and encryption, whereby the spin of an electron in a quantum dot represent a quantum bit or qubit of information.

Biologists are experimenting with composites of living cells and quantum dots. These could possibly be used to repair damaged neural pathways or to deliver drugs by activating the dots with light.

Once again, significant advances in manufacturing will be needed to realize the potential of quantum dots. For example, the quantum state needed to make a quantum computer is relatively easily to create, but its behaviour is still unpredictable.

Nano Tools and Fabrication Techniques

Legend has it that the people who profited most from the Klondike gold rush at the end of the 19th century were the sellers of picks and shovels. The same may hold true for nanotechnology, at least in the coming decade before production techniques are improved. According to market researchers Freedonia, the $245 million nanotech tools industry will grow by 30 % annually over the next few years. Microscopes and related tools dominate now, but measurement, fabrication/production and simulation/modelling tools will grow the fastest. Electronics and life sciences markets will emerge first; industrial, construction, energy generation and other applications will arise later.

Microscopy

Nanotechnology uses two main kinds of microscopy. The first involves a stationary sample in line with a high-speed electron gun. Both the *scanning electron microscope* (SEM) and *transmission electron microscope* (TEM) are based on this technique. The second class of microscopy involves a stationary scanner and a moving sample. The two microscopes in this class are the *atomic force microscope* (AFM) and the *scanning tunnelling microscope* (STM).

Microscopy plays a paradoxical role in nanotechnology because, although it is the key to understanding materials and processes, on a nanoscale samples can be damaged by the high-energy electrons fired at them. This is not a problem with STM, but a further drawback is that most microscopes require very stringent sample preparation. The SEM, TEM, and STM need well prepared samples that are also electrically conductive. There are ways to get around this, but the fact remains that it can take hours to prepare and mount a sample correctly (and hours to actually synthesize the sample).

Top-down and bottom-up synthesis techniques

There are two approaches to building nanostructures, both having their origins in the semiconductor industry. In the traditional "*top-down*" approach a larger material such as a silicon wafer is processed by removing matter until only the nanoscale features remain. Unfortunately, these techniques require the use of lithography, which requires a mask that selectively protects portions of the wafer from light. The distance from the mask to the wafer, and the size of the slit define the minimum feature size possible for a given frequency of light, e.g. extreme ultraviolet light yields feature sizes of 90 nm across, but this scale is near the fundamental limit of lithography. Nonetheless lithography can be used for patterning substrates used to produce nanomaterials, e.g. guiding the growth of quantum dots and nanowires.

The "*bottom-up*" approach starts with constituent materials (often gases or liquids) and uses chemical, electrical, or physical forces to build a nanomaterial atom-by-atom or molecule-by-molecule. The simplest bottom up synthesis route is *electroplating* to create a material layer-by-layer, atom-by-atom. By inducing an electric field with an applied voltage, charged particles are attracted to the surface of a substrate where bonding will occur. Most nanostructured metals with high hardness values are created with this approach. *Chemical vapour deposition* (CVD) uses a mix of volatile gases and takes advantage of thermodynamic principles to have the source material migrate to the substrate and then bond to the surface. This is the one proven method for creating nanowires and carbon nanotubes, and is a method of choice for creating quantum dots. Molecular self-assembly promises to be a revolutionary new way of creating materials from the bottom up. One way to achieve self-assembly is to use attractive forces like static electricity, Van der Waals forces, and a variety of other short-range forces to orient constituent molecules in a regular array. This has proven very effective in creating large grids of quantum dots.

The *bottom up* approach promises an unheard-of level of customisability in materials synthesis, but controlling the process is not easy and can only produce simple structures, in time-consuming processes with extremely low yields. It is not yet possible to produce integrated devices from the bottom up, and any overall order aside from repeating grids cannot be done without some sort of top-down influence like lithographic patterning. Nanotechnology synthesis is thus mainly academic, with only a few companies in the world that can claim to be nanotechnology manufacturers. And until understanding of synthesis is complete, it will be impossible to reach a point of mass production.

Present and Future Areas of Application

What is nanotechnology already used for?

Nanoscale materials, as mentioned above, have been used for many decades in several applications, are already present in a wide range of products, including mass-market consumer products. Among the most well-known are a glass for windows which is coated with *titanium oxide nanoparticles* that react to sunlight to break down dirt. When water hits the glass, it spreads evenly over the surface, instead of forming droplets, and runs off rapidly, taking the dirt with it. Nanotechnologies are used by the car industry to reinforce certain properties of car bumpers and to improve the adhesive properties of paints. Other uses of nanotechnologies in consumer products include:

Sunglasses using protective and anti-reflective ultrathin polymer coatings. Nanotechnology also offers scratch-resistant coatings based on nanocomposites that are transparent, ultra-thin, simple to care for, well suited for daily use and reasonably priced.

Textiles can incorporate nanotechnology to make practical improvements to such properties as wind-proofing and waterproofing, preventing wrinkling or staining, and guarding against electrostatic discharges. The wind-proof and waterproof properties of one *ski jacket*, for example, are obtained not by a surface coating of the jacket but by the use of nanofibers. Given that low-cost countries are capturing an ever-increasing share of clothes manufacturing, high-cost regions are likely to focus on high-tech clothes with the additional benefits for users that nanotech can help implement. Future projects include clothes with additional electronic functionalities, so-called "*smart clothes*" or "*wearable electronics*". These could include sensors to monitor body functions or release drugs in the required amounts, self-repairing mechanisms or access to the Internet.

Sports equipment manufacturers are also turning to nanotech. A high-performance *ski wax*, which produces a hard and fast-gliding surface, is already in use. The ultra-thin coating lasts longer than conventional waxing systems. Tennis rackets with carbon nanotubes have increased torsion and flex resistance. The rackets are more rigid than current carbon rackets and pack more power. Long-lasting tennis-balls are made by coating the inner core with clay polymer nanocomposites and have twice the lifetime of conventional balls.

Sunscreens and *cosmetics* based on nanotech are already widely used. Customers like products that are translucent because they suggest purity and cleanliness, and L'Oreal discovered that when lotions are ground down to 50 or 60 nms, they let light through. For sunscreens, mineral nanoparticles such as titanium dioxide offer several advantages. Traditional chemical UV protection suffers from its poor long-term stability. Titanium dioxide nanoparticles have a comparable UV protection property as the bulk material, but lose the cosmetically undesirable whitening as the particle size is decreased. For anti-wrinkle creams, a polymer capsule is used to transport active agents like vitamins.

Televisions using carbon nanotubes could be in use by late 2006 according to Samsung. Manufacturers expect these "field effect displays," (FED) to consume less energy than plasma or liquid crystal display (LCD) sets and combine the thinness of LCD and the image quality of traditional cathode ray tubes (CRT). The electrons in an FED are fired through a vacuum at a layer of phosphorescent glass covered with pixels. But unlike CRT, the electron source, the carbon, is only 1 to 2 mm from the target glass instead of 60 cm with CRT, and, instead of one electron source, the electron gun, there are thousands. FED contain less electronics than LCD and can be produced in a wide range of sizes. Toshiba, for example, will offer screen sizes of at least 50 inches, around 130 cm.

What applications are foreseen in the medium term?

The following list gives a quick overview of the many domains where nanotechnology is expected to fundamentally change products and how they are produced over the next two decades.

Electronics and communications: recording using nanolayers and dots, flat-panel displays, wireless technology, new devices and processes across the entire range of communication and information technologies, factors of thousands to millions improvements in both data storage capacity and processing speeds and at lower cost and improved power efficiency compared to present electronic circuits.

Chemicals and materials: catalysts that increase the energy efficiency of chemical plants and improve the combustion efficiency (thus lowering pollution emission) of motor vehicles, super-hard and tough (i.e., not brittle) drill bits and cutting tools, "*smart*" magnetic fluids for vacuum seals and lubricants.

Pharmaceuticals, health-care, and life sciences: nanostructured drugs, gene and drug delivery systems targeted to specific sites in the body, bio-compatible replacements for body parts and fluids, self-diagnostics for use in the home, sensors for labs-on-a-chip, material for bone and tissue regeneration.

Manufacturing: precision engineering based on new generations of microscopes and measuring techniques, new processes and tools to manipulate matter at an atomic level, nanopowders that are sintered into bulk materials with special properties that may include sensors to detect incipient failures and actuators to repair problems, chemical-mechanical polishing with nanoparticles, self-assembling of structures from molecules, bio-inspired materials and biostructures.

Energy technologies: new types of batteries, artificial photosynthesis for clean energy, quantum well solar cells, safe storage of hydrogen for use as a clean fuel, energy savings from using lighter materials and smaller circuits.

Space exploration: lightweight space vehicles, economic energy generation and management, ultrasmall and capable robotic systems.

Environment: selective membranes that can filter contaminants or even salt from water, nanostructured traps for removing pollutants from industrial effluents, characterization of the effects of nanostructures in the environment, maintenance of industrial sustainability by significant reductions in materials and energy use, reduced sources of pollution, increased opportunities for recycling.

National security: detectors and detoxifiers of chemical and biological agents, dramatically more capable electronic circuits, hard nanostructured coatings and materials, camouflage materials, light and self-repairing textiles, blood replacement, miniaturized surveillance systems.

Market Prospects and Opportunities

As an OECD survey emphasizes, most statistical offices do not collect data on nanotechnology R&D, human resources or industrial development, in part because nanotechnology remains a relatively new field of science and technology (and even more so of government policy)

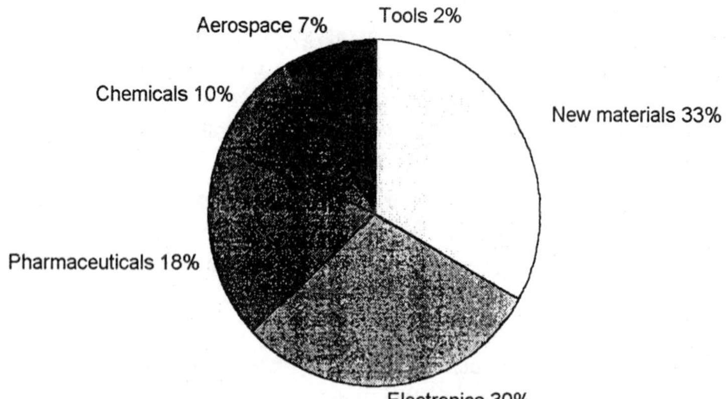

Fig. 4.1. Projected contribution of nanotechnology.

and also because of its interdisciplinary and cross-sectoral character. Given this, estimates of potential nanotech markets tend to come from private sources such as specialized consultancy firms who survey a wide number of actors in the field. Lux Research, for example, states that: "Sales of products incorporating emerging nanotechnology will rise from less than 0.1% of global manufacturing output today to 15% in 2014, totalling $2.6 trillion. This value will approach the size of the information technology and telecom industries combined and will be 10 times larger than biotechnology revenues". Insurers SwissRe echo this: "Sales revenues from products manufactured using nanotechnology have already reached eleven-digit figures and are projected to generate twelve-digit sums by 2010, even thirteen-digit sums by 2015". The chart below shows a projection for the US economy from the National Science Foundation.

These estimates should be treated with caution, because, as their authors point out, they refer to products "*incorporating nanotechnology*" or "*manufactured using nanotechnology*", not to nanotechnology products as such. (It's as if the value textile industry were calculated by including everything "*incorporating*" textiles, be it clothes, aircraft or automobiles.) Lux actually evaluates sales of basic nanomaterials like carbon nanotubes at $13 billion in 2014, a considerable sum, but far from $2.6 trillion. Although the accountancy may be contested, the projected three-phase growth path seems credible:

1. In the present phase, nanotechnology is incorporated selectively into high-end products, especially in automotive and aerospace applications.

2. Through 2009, commercial breakthroughs unlock markets for nanotechnology innovations. Electronics and IT applications dominate as microprocessors and memory chips built using new nanoscale processes come to market.
3. From 2010 onwards, nanotechnology becomes commonplace in manufactured goods. Health- care and life sciences applications finally become significant as nano-enabled pharmaceuticals and medical devices emerge from lengthy human trials.

Following the dotcom fiasco, potential investors are justifiably wary of treating nanotech as "the *next big thing*". While the excitement and type generated by nanotech's apostles may be reminiscent of how internet was going to change the world and make everybody rich, there are two crucial differences that counteract the likelihood of a nanobubble: nanotechnology is extremely difficult and extremely expensive, which is why it is concentrated in well-funded companies and institutes that can attract and nurture the scarce scientific and technical expertise needed to understand the problems and challenges. Moreover, the long lead times involved in moving from concept to commercialization make nanotech particularly unsuitable for making money fast.

Medicine

Medical and life-science applications may prove to be the most lucrative markets for nanotechnologies, with "*lab-on-a-chip*" devices already being manufactured and animal testing and early clinical trials starting on nanotechniques for drug delivery. However, the long product approval processes typical of the domain may mean that the health benefits to users and economic benefits to companies will take longer to realize than in other domains. Nanotech's promise comes from the fact that nanoscale devices are a hundred to ten thousand times smaller than human cells and are similar in size to large biological molecules ("biomolecules") such as enzymes and receptors. For example, haemoglobin, the molecule that carries oxygen in red blood cells, is approximately 5 nm in diameter, DNA 2.5, while a quantum dot is about the same size as a small protein (< 10 nm) and some viruses measure less than 100 nm. Devices smaller than 50 nm can easily enter most cells, while those smaller than 20 nm can move out of blood vessels as they circulate through the body.

Because of their small size, nanoscale devices can readily interact with biomolecules on both the surface of cells and inside of cells. By gaining access to so many areas of the body, they have the potential to detect disease and deliver treatment in new ways. Nanotechnology offers the opportunity to study and interact with cells at the molecular and cellular scales in real time, and during the earliest stages of the development of a disease. And since nanocomponents can be made to share some of the same properties as natural nanoscale structures, it is hoped to develop artificial nanostructures that sense and repair damage to the organism, just as naturally-occurring biological nanostructures such as white blood cells do.

Cancer research illustrates many of the medical potentials of nanotechnologies in the longer term. It is hoped that nanoscale devices and processes will help to develop:

1. Imaging agents and diagnostics that will allow clinicians to detect cancer in its earliest stages;
2. Systems that will provide real-time assessments of therapeutic and surgical efficacy for accelerating clinical translation;
3. Multifunctional, targeted devices capable of bypassing biological barriers to deliver multiple therapeutic agents directly to cancer cells and those tissues in the microenvironment that play a critical role in the growth and metastasis of cancer,
4. Agents that can monitor predictive molecular changes and prevent precancerous cells from becoming malignant,
5. Novel methods to manage the symptoms of cancer that adversely impact quality of life,
6. Research tools that will enable rapid identification of new targets for clinical development and predict drug resistance.

Drug delivery

This may be the most profitable application of nanotechnology in medicine, and even generally, over the next two decades. Drugs need to be protected during their transit through the body to the target, to maintain their biological and chemicals properties or to stop them damaging the parts of the body they travel through. Once a drug arrives at its destination, it needs to be released at an appropriate rate for it to be effective. This process is called encapsulation, and nanotechnology can improve both the diffusion and degradation characteristics of the encapsulation material, allowing the drug to travel efficiently to the target and be released in an optimal way. Nanoparticle encapsulation is also being investigated for the treatment of neurological disorders to deliver therapeutic molecules directly to the central nervous system beyond the blood-brain barrier, and to the eye beyond the blood-retina barrier. Applications could include Parkinson's, Huntington's chorea, Alzheimer's, ALS and diseases of the eye.

Repair and replacement

Damaged tissues and organs are often replaced by artificial substitutes, and nanotechnology offers a range of new biocompatible coatings for the implants that improves their adhesion, durability and life-span. New types of nanomaterials are being evaluated as implant coatings to improve interface properties. For example, nanopolymers can be used to coat devices in contact with blood (e.g. artificial hearts, catheters) to disperse clots or prevent their formation. Nanomaterials and nanotechnology fabrication techniques are being investigated as tissue regeneration scaffolds. The ultimate goal is to grow large complex organs. Examples include nanoscale polymers moulded into heart valves, and polymer nanocomposites for bone scaffolds. Commercially viable solutions are thought to be 5 to 10 years away, given the scientific challenges related to a better understanding of molecular/cell biology and fabrication methods for producing large three-dimensional scaffolds.

Nanostructures are promising for temporary implants, e.g. that biodegrade and do not have to be removed in a subsequent operation. Research is also being done on a flexible nanofiber membrane mesh that can be applied to heart tissue in open-heart surgery. The mesh can be infused with antibiotics, painkillers and medicines in small quantities and directly applied to internal tissues.

Subcutaneous chips are already being developed to continuously monitor key body parameters including pulse, temperature and blood glucose. Another application uses optical microsensors implanted into subdermal or deep tissue to monitor tissue circulation after surgery, while a third type of sensor uses MEMS (microelectromechanical system) devices and accelerometers to measure strain, acceleration, angular rate and related parameters for monitoring and treating paralysed limbs, and to improve the design of artificial limbs. Implantable sensors can also work with devices that administer treatment automatically if required, e.g. fluid injection systems to dispense drugs. Initial applications may include chemotherapy that directly targets tumours in the colon and are programmed to dispense precise amounts of medication at convenient times, such as after a patient has fallen asleep. Sensors that monitor the heart's activity level can also work with an implantable defibrillator to regulate heartbeats.

Hearing and vision

Nano and related micro technologies are being used to develop a new generation of smaller and potentially more powerful devices to restore lost vision and hearing. One approach uses a miniature video camera attached to a blind person's glasses to capture visual signals processed by a microcomputer worn on the belt and transmitted to an array of electrodes placed in the eye. Another approach uses of a subretinal implant designed to replace photoreceptors in the retina. The implant uses a microelectrode array powered by up to 3500 microscopic solar cells.

For hearing, an implanted transducer is pressure-fitted onto a bone in the inner ear, causing the bones to vibrate and move the fluid in the inner ear, which stimulates the auditory nerve. An array at the tip of the device uses up to 128 electrodes, five times higher than current devices, to simulate a fuller range of sounds. The implant is connected to a small microprocessor and a microphone in a wearable device that clips behind the ear. This captures and translates sounds into electric pulses transmitted by wire through a tiny hole made in the middle ear.

Food and Agriculture

Nanotechnology is rapidly converging with biotech and information technology to radically change food and agricultural systems. Over the next two decades, the impacts of nanoscale convergence on farmers and food could even exceed that of farm mechanization or of the Green Revolution according to some sources such as the ETC group. Food and nutrition products containing nanoscale additives are already commercially available. Likewise, a number of pesticides formulated at the nanoscale are on the market and have been released in the environment. According to Helmut Kaiser Consultancy, some 200 transnational food companies are currently investing in nanotech and are on their way to commercializing products. The US leads, followed by Japan and China. HKC expects the nanofood market to surge from $2.6 billion in 2003 to $7.0 billion in 2006 and to $20.4 billion in 2010.

Companies not associated with food production in the public mind are already supplying nano-enabled ingredients to the industry. BASF, for example, exploits the fact that many vitamins and other substances such as carotinoids are insoluble in water, but can easily be mixed with cold water when formulated as nanoparticles. Many lemonades and fruit juices contain these specially formulated additives, which can also be used to provide an "*attractive*" colour.

Expected breakthroughs in crop *DNA decoding* and analysis could enable agrifirms to predict, control and improve agricultural production. And with technology for manipulating the molecules and atoms of food, the food industry would have a powerful method to design food with much greater capability and precision, lower costs and improved sustainability. The combination of DNA and nanotechnology research could also generate new nutrition delivery systems, to bring active agents more precisely and efficiently to the desired parts of the human body.

Nanotechnology will not only change how every step of the food chain operates but also who is involved. At stake is the world's $3 trillion food retail market, agricultural export markets valued at $544 billion, the livelihoods of farmers and the well-being of the rest of us. Converging technologies could reinvigorate the battered agrochemical and agbiotech industries, possibly igniting a still more intense debate – this time over "*atomically-modified*" foods.

The most cited nano-agricultural developments are:

Nanoseeds

In Thailand, scientists at Chiang Mai University's nuclear physics laboratory have rearranged the DNA of rice by drilling a nano-sized hole through the rice cell's wall and membrane and inserting a nitrogen atom. So far, they've been able to change the colour of the grain, from purple to green.

Nanoparticle pesticides

Monsanto, Syngenta and BASF are developing pesticides enclosed in nanocapsules or made up of nanoparticles. The pesticides can be more easily taken up by plants if they're in nanoparticle form; they can also be programmed to be "*time-released.*"

Nanofeed for chickens

With funding from the US Department of Agriculture (USDA), Clemson University researchers are feeding bioactive polystyrene nanoparticles that bind with bacteria to chickens as an alternative to chemical antibiotics in industrial chicken production.

Nano ponds

One of the USA's biggest farmed fish companies, Clear Spring Trout, is adding nanoparticle vaccines to trout ponds, where they are taken up by fish.

"Little brother"

The USDA is pursuing a project to cover farmers' fields and herds with small wireless sensors to replace farm labour and expertise with a ubiquitous surveillance system.

Nano foods

Kraft, Nestle, Unilever and others are employing nanotech to change the structure of food—creating "interactive" drinks containing nanocapsules that can change colour and flavour (Kraft) and spreads and ice creams with nanoparticle emulsions to improve texture. Others are inventing small nanocapsules that will smuggle nutrients and flavours into the body (what one company calls "*nanoceuticals*").

Nano packaging

BASF, Kraft and others are developing new nanomaterials that extend food shelf life and signal when a food spoils by changing colour.

Food safety

Scientists from the University of Wisconsin have successfully used single bacterial cells to make tiny bio-electronic circuits, which could in the future be used to detect bacteria, toxins and proteins.

Nanosensors can work through a variety of methods such as by the use of nanoparticles tailor-made to fluoresce different colours or made from magnetic materials can selectively attach themselves to food pathogens. Handheld sensors employing either infrared light or magnetic materials could then note the presence of even minuscule traces of harmful pathogens. The advantage of such a system is that literally hundreds and potentially thousands of nanoparticles can be placed on a single nanosensor to rapidly, accurately and affordably detect the presence of any number of different bacteria and pathogens. A second advantage of nanosensors is that given their small size they can gain access into the tiny crevices where the pathogens often hide, and nanotechnology may reduce the time it takes to detect the presence of microbial pathogens from two to seven days down to a few hours and, ultimately, minutes or even seconds.

Semiconductors and Computing

The computer industry is already working on a nanoscale. Although the current production range is at 90 nm, 5 nm gates have been proven in labs, although they cannot be manufactured yet. By 2010, worldwide, about $300 billion worth of semiconductor production will be nanotechnology-based (including nanocomponents such as nanolayers, nanoscale treated materials, or other nanostructures) and by 2015, about

$500 billion. Because nanotechnology can reduce its basic features, CMOS will continue being used for a decade or more. The intermediate future will have CMOS married to a generation of nanodevices as yet undefined, because there are many alternatives, and it is still too early to tell which will prevail. One solution could be hybrid structures exploiting the advantages of today's CMOS technology (integration and scaling of transistors and high functionality on a small support) with off-chip optoelectronic interconnects to overcome the throughput bottlenecks.

Towards 2015, semiconductor development priorities will change, as the focus shifts from scaling and speed to system architecture and integration, with user-specific applications for bio-nanodevices, the food industry and construction applications. Another trend is the convergence between IT, nanotechnology, biotechnology and cognitive sciences. The higher speeds at which information will be disseminated will change how we work with computers, and also perhaps how we deal with things like damaged nerves, possibly by developing direct interfaces with the nervous system and electronic circuits, so-called neuromorphic engineering, where signals are directly transmitted from a human organism to a machine.

The actual technologies employed are hard to predict. Currently there exist at least four interrelated technical barriers to nanoscale manufacturing: How to control the assembly of 3-D heterogeneous systems, including alignment, registration and interconnection at 3-D and with multiple functionalities.

How to handle and process nanoscale structures in a high-rate/ high-volume manner without compromising beneficial nanoscale properties.

How to test nanocomponents' long-term reliability, and detect, remove or prevent defects and contamination. Metrology. At present, using an electron microscope, it is possible to get depth of field, sufficient resolution or low energy (important so as not to damage certain components), but not all three at once. Failure analysis is another metrology issue: how to get a real 3-D view of the structure and defects that may develop during processing or use.

At present, technology front runners include *spin electronics*, *molecular electronics*, *biocomponents*, *quantum computing*, *DNA computing*, etc. However, the history of technology teaches that sudden upsets that could change everything are to be expected. As recently as 1998, limited use was predicted for giant magnetoresistance introduced by IBM. But within two years it replaced all equivalent hard disk reading technologies and their extensive production facilities. The technique exploits the electron's spin to produce novel interconnect and device structures, giving rise to the name "*spintronics*". Spin is present in all electrons, and manipulating spin would use conventional solid-state semiconductor and metal materials, without the problems associated with nanotubes or molecules. Spin packets have a long lifetime and high mobility in semiconductors, making them attractive for transmitting information in the chip, within the silicon, without using a metal. One major problem with spintronics is that when a magnet heats up, it ceases being ferromagnetic, a condition necessary to exploit the electron spin. It is also difficult to control the ferromagnetic force or direction.

Assuming these problems can be solved, promising applications for spintronics include MRAM (magnetic random access memory), a high-speed nonvolatile memory architecture; and logic devices like the spin field effect transistor (spin FET), which consumes less power and operates faster than its conventional counterpart.

Chip makers are already working at around 100 nm, but this is essentially a "*shrinking*" of conventional technologies to make them smaller, and this is now reaching its limits. Miniaturization to much smaller scales will run into problems caused by quantum phenomena, such as electrons tunnelling through the barriers between wires, so an alternative to transistor technology must be found, one whose components will exploit quantum effects rather suffer from them. The first generation of nanocomputers will have components that behave according to quantum mechanics, but their algorithms will probably not involve quantum mechanics. If quantum mechanics could be used in the algorithms as well, the computer would be enormously more powerful than any classical scheme, but such developments are unlikely in the foreseeable future.

In the meantime, research is in progress to manipulate molecules to carry out calculations. In "chemical computing", a series of chemical reactions, e.g. of DNA, corresponds to a computation, with the final products of the reactions representing the answer. With this technique, many calculations can be carried out in parallel, but each step requires a long time, and can be very expensive because of the cost of the chemicals used.

A second approach is to use molecules as the "host" for nuclear spins that form the quantum bits (qubits) in a nuclear magnetic resonance-based computer. However, this approach may not be able to scale up to a computationally useful number of qubits.

The most promising approach is thought to be molecular electronics, using a molecule or group of molecules in a circuit. Bit densities for molecular logic and memory components could be on the order of a terabit/cm^2 (6.5 terabits/in^2). Switching speeds could get down into the range of a few picoseconds (1000 times faster than current DRAM).

Textiles

The textile industry could be affected quite significantly by nanotechnology, with some estimates talking of a market impact of hundreds of billions of dollars over the next decade. Nanoscience has already produced stain- and wrinkle-resistant clothing, and future developments will focus on upgrading existing functions and performances of textile materials; and developing "smart" textiles with unprecedented functions such as: (i) sensors and information acquisition and transfer, (ii) multiple and sophisticated protection and detection, (iii) health-care and wound-healing functions, and (iv) self-cleaning and repair functions.

This last function illustrates how nanotechnology could impact areas outside its immediate application.

US company *Nano-Tex* is already marketing its *NanoCare stain-* and wrinkle-resistant technology, and *NanoFresh* (to freshen sports clothing) is expected soon. Scientists at the Hong Kong Polytechnic University have built a nano layer of particles of titanium dioxide, a substance that reacts with sunlight to break down dirt and other organic material. This layer can be coated on cotton to keep the fabric clean. Clothes simply need to be exposed to natural or ultraviolet light for the cleaning process to begin. Once triggered by sunlight, clothing made out of the fabric will be able to rid itself of dirt, pollutants and microorganisms. The whole laundry industry would be affected if the technology proves to be economically viable.

Research involving nanotechnology to improve performances or to create new functions is most advanced in nanostructured composite fibers employing nanosize fillers such as nanoparticles (clay, metal oxides, carbon black), graphite nanofibers (GNF) and carbon nanotubes (CNT). The main function of nanosize fillers is to increase mechanical strength and improve physical properties such as conductivity and antistatic behaviours. Being evenly distributed in polymer matrices, nanoparticles can carry load and increase the toughness and abrasion resistance; nanofibers can transfer stress away from polymer matrices and enhance tensile strength of composite fibers. Additional physical and chemical performances imparted to composite fibers vary with specific properties of the nanofillers used. Although some of the filler particles such as clay, metal oxides, and carbon black have previously been used as microfillers in composite materials for decades, reducing their size into nanometer range have resulted in higher performances and generated new market interest.

Carbon nanofibers and carbon nanoparticles

Carbon nanofibers and *carbon black nanoparticles* are among the most commonly used nanosize filling materials. Carbon nanofibers can effectively increase the tensile strength of composite fibers due to their high aspect ratio, while carbon black nanoparticles can improve abrasion resistance and toughness. Both have high chemical resistance and electric conductivity.

Clay nanoparticles

Clay nanoparticles or nanoflakes possess electrical, heat and chemical resistance and an ability to block UV light. Composite fibers reinforced with clay nanoparticles exhibit flame retardant, anti-UV and anticorrosive behaviours.

Metal oxide nanoparticles

Certain metal oxide *nanoparticles* possess photocatalytic ability, electrical conductivity, UV absorption and photooxidizing capacity against chemical and biological species. Research involving these nanoparticles focuses on antimicrobial, self-decontaminating and UV blocking functions for both military protection gear and civilian health products.

Carbon nanotubes

Potential applications of CNTs include conductive and high-strength composite fibers, energy storage and energy conversion devices, sensors, and field emission displays. One CNT fiber already exhibits twice the stiffness and strength, and 20 times the toughness of steel wire of the same weight and length. Moreover, toughness can be four times higher than that of spider silk and 17 times greater than Kevlar fibers used in bulletproof vests, suggesting applications in safety harnesses, explosion-proof blankets, and electromagnetic shielding.

Nanotechnology in textile finishing

Nanoscale emulsification, through which finishes can be applied to textile material in a more thorough, even and precise manner provide an unprecedented level of textile performance regarding stain-resistant, hydrophilic, antistatic, wrinkle resistant and shrinkproof properties.

Nanosize metal oxide and ceramic particles have a larger surface area and hence higher efficiency than larger size particles, are transparent, and do not blur the colour and brightness of the textile substrates. Fabric treated with nanoparticles $Ti0_2$ and MgO replaces fabrics with active carbon, previously used as chemical and biological protective materials. The photocatalytic activity of $Ti0_2$ and MgO nanoparticles can break down harmful chemicals and biological agents.

Finishing with nanoparticles can convert fabrics into sensor-based materials. If nanocrystalline piezoceramic particles are incorporated into fabrics, the finished fabric can convert exerted mechanical forces into electrical signals enabling the monitoring of bodily functions such as heart rhythm and pulse if they are worn next to skin.

Self assembled nanolayers

In the longer-term future, *self-assembled nanolayer* (*SAN*) coating may challenge traditional textile coating. Research in this area is still in the very early stages, but the idea is to deposit a coating less than one nanometer thick on the textile, and then to vary the number of successive nanolayers to modulate the desired physical properties of the finished article.

Energy

Breakthroughs in nanotechnology could provide technologies that would contribute to worldwide energy security and supply. A report published by Rice University (Texas) in February 2005 identified numerous areas in which nanotechnology could contribute to more efficient, inexpensive, and environmentally sound technologies than are readily available. Although the most significant contributions may be to unglamorous applications such as better materials for exploration equipment used in the oil and gas industry or improved catalysis, nanotechnology is being proposed in numerous energy domains, including solar power; wind; clean coal; fusion reactors; new generation fission reactors; fuel cells; batteries; hydrogen production, storage and transportation; and a new electrical grid that ties all the power sources together. The main challenges where nanotechnology could contribute are:

1. Lower the costs of photovoltaic solar energy tenfold.
2. Achieve commercial photocatalytic reduction of CO_2 to methanol.
3. Create a commercial process for direct photoconversion of light and water to produce hydrogen.

4. Lower the costs of fuel cells between tenfold and a hundredfold and create new, sturdier materials.
5. Improve the efficiency and storage capacity of batteries and supercapacitors between tenfold and a hundredfold for automotive and distributed generation applications.
6. Create new lightweight materials for hydrogen storage for pressure tanks, liquid hydrogen vessels, and an easily reversible hydrogen chemisorption system.
7. Develop power cables, superconductors or quantum conductors made of new nanomaterials to rewire the electricity grid and enable long-distance, continental and even international electrical energy transport, also reducing or eliminating thermal sag failures, eddy current losses and resistive losses by replacing copper and aluminium wires.
8. Develop thermochemical processes with catalysts to generate hydrogen from water at temperatures lower than 900C at commercial costs.
9. Create superstrong, lightweight materials that can be used to improve energy efficiency in cars, planes and in space travel; the latter, if combined with nanoelectronics based robotics, possibly enabling space solar structures on the moon or in space.
10. Create efficient lighting to replace incandescent and fluorescent lights.
11. Develop nanomaterials and coatings that will enable deep drilling at lower costs to tap energy resources, including geothermal heat, in deep strata.
12. Create CO_2 mineralization methods that can work on a vast scale without waste streams.

Solving these challenges will take many years, but commercial and public research institutes are already exploiting nanotechnology for energy applications. Bell Labs, for example, is exploring the possibility of producing a microbattery that would still work 20 years after purchase by postponing the chemical reactions that degrades traditional batteries. The battery is based on a Bell Labs discovery that liquid droplets of electrolyte will stay in a dormant state atop microscopic structures called "nanograss" until stimulated to flow, thereby triggering a reaction producing electricity. Other researchers hope to dispense with batteries completely by developing nanotubes-based "*ultra*" capacitors powerful enough to propel hybrid-electric cars.

Compared with batteries, ultracapacitors can put out much more power for a given weight, can be charged in seconds rather than hours, and can function at more extreme temperatures. They're also more efficient, and they last much longer. The technology is in it earlier stages: the worldwide market was only $38 million in 2002, the most recent year for which figures are available, but researcher firm Frost & Sullivan predicts total 2007 revenues for ultracapacitors of $355 million.

Photovoltaics is another area where nanotech is already providing products that could have a significant impact. Three US-based solar cell start-ups (Nanosolar, Nanosys and Konarka Technologies), and corporate players including Matsushita and STMicroelectronics are striving to produce photon-harvesting materials at lower costs and in higher volumes than traditional crystalline silicon photovoltaic cells. Nanosolar has developed a material of metal oxide nanowires that can be sprayed as a liquid onto a plastic substrate where it self-assembles into a photovoltaic film. A roll-to-roll process similar to high-speed printing offers a high-volume approach that does not require high temperatures or vacuum equipment. Nanosys intends its solar coatings to be sprayed onto roofing tiles. And Konarka is developing plastic sheets embedded with titanium dioxide nanocrystals coated with light-absorbing dyes. The company acquired Siemens' organic photovoltaic research activities, and Konarka's recent $18 million third round of funding included the world's first- and fifth-largest energy companies, Electricite de France and Chevron-Texaco. If nanotech solar fabrics could be applied to, e.g., buildings and bridges, the energy landscape could change in important ways. Integrated into the roof of a bus or truck, they could split water via electrolysis and generate hydrogen to run a fuel cell. Losers would include current photovoltaic-cell makers and battery manufacturers who failed to react to the new challenge.

Such developments however depend on solving a number of fundamental problems at the nanoscale, but researchers are making fast progress using nanoscale design, include accelerating the kinetics of reactions through catalysis, separating the products at high temperature, and directing products to the next reaction step.

Nanotechnology and the Situation of Developing Countries

While research and development in nanotechnology is quite limited in most developing countries, there will be increasing opportunities to import nano products and processes. It can be argued of course that nanotechnology could make the situation of developing countries worse by reducing demand for their exports, notably raw materials. Moreover, even in developing countries, few nanotech projects specifically target the needs of the poor, leading to fears of a "nano divide" similar to the digital divide. The UN's International Centre for Science and High Technology tackled such issues in February 2005 at a meeting entitled "North- South dialogue on nanotechnology". The Centre argued that nanotechnology may offer important benefits to developing countries and it is not correct to assume that it is too difficult or too expensive for them. A similar theme was the subject of a report published in April 2005 by the Canadian Program on Genomics and Global Health (CPGGH) at the University of Toronto Joint Centre for Bioethics (JCB). The CPGGH asked over 60 international experts to assess the potential impacts of nanotechnologies for developing countries within the framework of the UN Millennium Development Goals. Agreed in 2000 for achievement by 2015, the UN goals are: to halve extreme poverty and hunger; achieve universal primary education; empower women and promote equality between women and men; reduce under-five mortality by two-thirds; reduce maternal mortality by three-quarters; reverse the spread of diseases, especially HIV/AIDS and malaria; ensure environmental sustainability; and create a global partnership for development.

The CPGGH study ranks the 10 nanotechnology applications most likely to have an impact in the areas of water, agriculture, nutrition, health, energy and the environment by 2015. The ranking is markedly different from similar exercises in the more advanced industrial economies, where applications in electronics and computing are generally seen as the most significant, with pharmaceuticals and other health sectors featuring strongly. For developing countries, the experts reckon the top 10 nanotechnology applications are:

Energy

There was a high degree of unanimity in ranking this area number 1. Nanomaterials are being used to build a new generation of solar cells, hydrogen fuel cells and novel hydrogen storage systems that could deliver clean energy to countries still reliant on traditional, non-renewable contaminating fuels. Advances in the creation of synthetic nanomembranes embedded with proteins are capable of turning light into chemical energy. If successfully developed on an industrial scale, such technologies could help developing countries avoid recurrent shortages and price fluctuations that come with dependence on fossil fuels, as well as the environmental consequences of mining and burning oil and coal.

Agriculture

Researchers are developing a range of inexpensive nanotech applications to increase soil fertility and crop production, and help eliminate malnutrition—a contributor to more than half the deaths of children under five in developing countries. Nanotech materials are in development for the slow release and efficient dosage of fertilizers for plants and of nutrients and medicines for livestock. Other agricultural developments include nanosensors to monitor the health of crops and farm animals and magnetic nanoparticles to remove soil contaminants.

Water treatment

Nano-membranes and nano-clays are inexpensive, portable and easily cleaned systems that purify, detoxify and desalinate water more efficiently than conventional bacterial and viral filters. Researchers also have developed a method of large-scale production of carbon nanotube filters for water quality improvement. Other water applications include systems (based on titanium dioxide and on magnetic nanoparticles) that decompose organic pollutants and remove salts and heavy metals from liquids, enabling the use of heavily contaminated and salt water for irrigation and drinking. Several of the contaminating substances retrieved could then be easily recycled.

Disease diagnosis and screening

Technologies include the "lab-on-a-chip", which offers all the diagnostic functions of a medical laboratory, and other biosensors based on nanotubes, wires, magnetic particles and semiconductor crystals (quantum dots). These inexpensive, hand-held diagnostic kits detect the presence of several pathogens at once and could be used for wide-range screening in small peripheral clinics. Other nanotechnology applications are in development that would greatly enhance medical imaging.

Drug delivery systems

Nanocapsules, dendrimers (tiny bush-like spheres made of branched polymers), and "buckyballs" (soccerball-shaped structures made of 60 carbon atoms) for slow, sustained drug release systems, characteristics valuable for countries without adequate drug storage capabilities and distribution networks. Nanotechnology could also potentially reduce transportation costs and even required dosages by improving shelf-life, thermostability and resistance to changes in humidity of existing medications;

Food processing and storage

Improved plastic film coatings for food packaging and storage may enable a wider and more efficient distribution of food products to remote areas in less industrialized countries; antimicrobial emulsions made with nanomaterials for the decontamination of food equipment, packaging, or food; and nanotech-based sensors to detect and identify contamination;

Air pollution remediation

Nanotech-based innovations that destroy air pollutants with light; make catalytic converters more efficient, cheaper and better controlled; detect toxic materials and leaks; reduce fossil fuel emissions; and separate gases.

Construction

Nano-molecular structures to make asphalt and concrete more resistant to water; materials to block ultraviolet and infrared radiation; materials for cheaper and durable housing, surfaces, coatings, glues, concrete, and heat and light exclusion; and self-cleaning for windows, mirrors and toilets.

Health monitoring

Nanodevices are being developed to keep track of daily changes in physiological variables such as the levels of glucose, of carbon dioxide, and of cholesterol, without the need for drawing blood in a hospital setting. For example, patients suffering from diabetes would know at any given time the concentration of sugar in their blood; similarly, patients with heart diseases would be able to monitor their cholesterol levels constantly.

Disease vector and pest detection control

Nanoscale sensors for pest detection, and improved pesticides, insecticides, and insect repellents.

Players

As the table below shows, most nanotech companies are in the US, mainly because of the more developed venture capital market (over half the venture capital investors in nanotechnology are from the US). Statistics for universities and research institutes also shows a strong, but less marked, US bias.

These figures should be treated with caution, since new companies are created all the time and some do not remain active very long. Moreover, the lack of any clear nanotech strategy even in companies

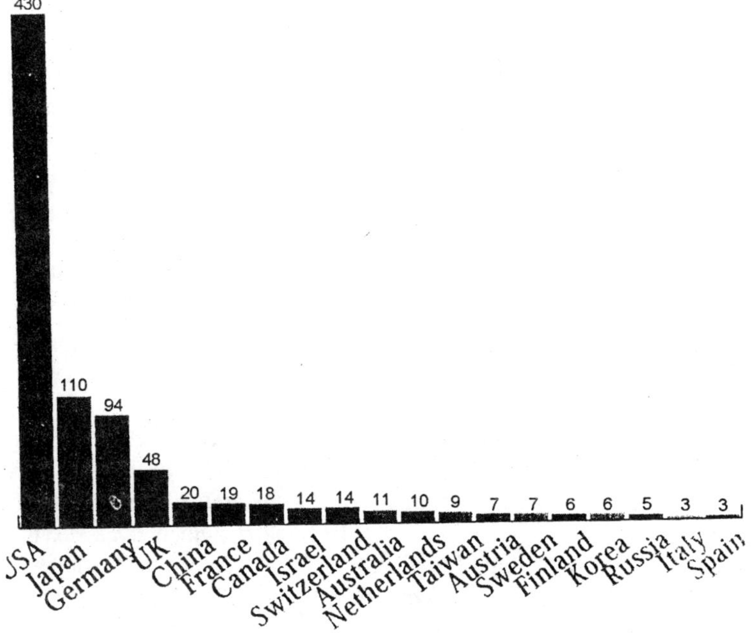

heavily involved in nanotechnology means that tomorrow's winners will not necessarily be today's heavy investors. A survey of executives responsible for nanotech at 33 companies with over $5 billion annual revenue found that they have not converged towards any particular model for organizing and governing their nanotechnology initiatives. Forty-two percent of represented companies have centralized nanotechnology programs, but an equal share pursue decentralized activity with no coordination. At 45% of companies, the R&D organization "owns" nanotech; ownership varies widely across the rest.

Barely half of interviews' firms have a stated nanotech strategy today. When a strategy does exist, it is frequently a platitude like "survey the field and move quickly." Fewer than half of interviews rate their companies' current approaches to nanotech as "very effective." Pharmaceutical companies are least likely to have an explicit nanotechnology strategy; they also invest the lowest level of people and funding compared with other sectors. Asian companies across industries show the highest levels of staffing, funding, and executive sponsorship for nanotech.

The roles of the different players in the nanotech sector are no different from those elsewhere:

1. Large organizations, with the resources to investigate longer-term technologies, seek applications to improve margins, lower costs or increase market share.
2. Start-ups, seeking to apply technologies in order to capture market share or disrupt existing markets, attract the attention of acquisition-hungry incumbents.
3. Economic blocs compete for supremacy, mindful of the economic benefits that strength in many of the applications of nanotechnology will bring.
4. Public agencies attempt to capture the maximum number of links in the value chain.

The long-term outcome is also likely to be similar, with, as happened in microcomputing, one or two new large companies emerging and most of the other viable start-ups being absorbed by large firms. Small companies are likely to target emerging technologies, seeking new generations of products, of which four generations were identified in 2004 by Mihail Roco, Chairman of the Federal Subcommittee on Nanoscale Science Engineering and Technology which oversees the National Nanotechnology Initiative. At the University of Southern California's "*Nano Ethics Conference*" in March 2005, Dr Roco revised his estimates for third and fourth generations and added even a fifth generation.

1. 2001. Passive nanostructures. This corresponds to the current state of affairs, with the creation of commercial prototypes and the acquisition of systematic control at the nanoscale for these products using nanostructure polymers, wires, coatings, etc,
2. 2005. "Active" nanostructures, i.e. devices such as actuators that behave like muscles; transistors with active parts created by design; drug delivery within the human body at specific locations and times. Such applications are already in advanced R&D, and some commercial prototypes are expected in the next year or so. Active nanostructure devices will lead to a significant market expansion,
3. 2015-20 (2010 in previous estimate). The third generation will arrive when nanodevices and nanomaterials are integrated into larger nanosystems, and systems of nanosystems with emerging behaviour will be created as commercial prototypes. This includes directed multiscale self-assembling and chemico-mechanical processing. By 2015 nanoscale designed catalysts will expand the use in "exact" chemical manufacturing to cut and link molecular assemblies, with minimal waste,

4. 2020 (previously 2015). Fourth-generation large nanosystems whose different components will be molecules or macromolecules will emerge over 2005- 2020. These will approximate how living systems work (except that living systems are more complex, being integrated on lengthier scales, generally use water, and grow slowly). In detail, this generation includes heterogeneous molecular nanosystems, where each molecule in the nanosystem has a specific structure and plays a different role. Molecules will be used as devices and from their engineered structures and architectures will emerge fundamentally new functions. Research focus will be on atomic manipulation for design of molecules and supramolecular systems, dynamics of single molecule, molecular machines, design of large heterogeneous molecular systems, controlled interaction between light and matter with relevance to energy conversion among others, exploiting quantum control, emerging behaviour of complex macromolecular assemblies, nanosystem biology for health-care and agricultural systems, human-machine interface at the tissue and nervous system level, and convergence of nano-bio-info-cognitive domains. Examples are creating multifunctional molecules, catalysts for synthesis and controlling of engineered nanostructures, subcellular interventions, and biomimetics for complex system dynamics and control,
5. Starting around 2030, with nanorobotics, guided assembly, and diverging architectures.

Risks of Nanotechnology

Variety of Risks

Keeping in mind the broad range of applications of nanotechnologies outlined in the previous chapters and the variety of industrial sectors that are affected, it is self-evident, that the risks associated with nanotechnologies will also form a complex risk landscape rather than a homogenous set of risks. The emphasis on what kind of risks are key to consider will depend on the perspective of the particular organization involved in nanotechnologies. To name just a few:

1. Business risks involved with marketing of nanotechnology enabled products.
2. Risks related to the protection of intellectual property.
3. Political risks regarding the impact on the economical development of countries and regions.
4. Risks regarding privacy when miniature sensors become ubiquitous.

5. Environmental risks from the release of nanoparticles into the environment.
6. Safety risks from nanoparticles for workers and consumers.
7. Futuristic risks like human enhancement and self replications of nano machines.

The catch-all term "*nanotechnology*" is not sufficiently precise for risk governance and risk management purposes. From a risk-control point of view it will be necessary to systematically identify those critical issues, which should be looked at in more detail. This risk identification process is a task for all parties involved and it should remain a dynamic process which always takes into account new scientific, technological, societal and legal trends.

This report mainly focuses on potential risks that are relevant to property and casualty insurance, rather than try to cover the broad range of topics indicated above.

Almost all safety concerns that have been raised about nanotechnologies are related to "*free*" rather than fixed engineered nanoparticles. The state of the discussion on these risks will be discussed.

The risk and safety discussion related to free *nanoparticles* will concern only a fraction of the applications of nanotechnologies. In most applications nanoparticles will be embedded in the final product and therefore not come into direct contact with workers, consumers or the environment. They are unlikely to raise concerns because of their immobilization. Exceptions are possible when the products or materials within which nanoparticles are enclosed are discarded, burned or otherwise destroyed (e.g. in an accident).

Looking at the manufacturing processes involved, they are similar to conventional well-established chemical and engineering processes. Building on established previous experience can be a guideline for risk assessment purposes, but the approach to health and safety issues needs to be modified to address the special characteristics of nanoparticles.

In the early phase of commercialization most applications of nanotechnology will improve existing products, rather than generating new products. In themselves, most areas of nanotechnology are not likely to present novel safety risks. As with all new technologies new risks might appear that we have not thought about yet, underlining the need for continuous and dynamic risk reviews.

Positive Effects on Human Health

A fair assessment of the risks of any new technology must also consider positive contributions to increased safety. The basic innovations that come from nanotechnologies have the potential to contribute to human health and environmental safety in many ways. They have the potential to contribute to solve urgent issues like the provision of clean drinking water or more efficient energy conversion and energy storage. The potential of nanotechnologies regarding economic benefits, the potential to create jobs, wealth and well being is very high. At the moment, public awareness about nanotechnology is limited. What happens over the next few years will determine how the public comes to view it. A transparent discussion of benefits and risks will help people reach a considered, balanced view. This will enable a greater public acceptance, which, in turn, will enable society as a whole to profit from these fundamental technological developments while, at the same time, the risks are kept under control.

Especially in the field of medicine there are quite a few technological developments that promise enhanced diagnostic possibilities, new ways to monitor patients, new ways to treat diseases like cancer and to reduce side effects. To give a few exampıes:

1. *Nanoparticles* can be used as carriers for targeted drug delivery. Their ability to penetrate certain protective membranes in the body, such as the blood-brain barrier, can be beneficial for many drugs. This could open the way for new drugs from active substances that have not been able to pass clinical trials due to less precise delivery mechanisms.
2. *Nanosensors* and lab-on-a-chip-technologies will foster early recognition and identification of diseases and can be used for continuous monitoring of patients with chronic diseases,
3. New therapeutic methods for the treatment of cancer with the help of nanoparticles are investigated.

Ultrasensitive detection of substances will have implications for safety in many other areas such as industrial medicine, environmental medicine and food safety. To give one example: it has ıecently been shown that bacterial pathogens can be detected in very low concentrations with the help of nanoparticles. Quick and accurate testing is crucial for avoiding potential infections, but in order to be effective many current tests require time-consuming amplification of samples. The new methods are very powerful: new findings indicate that specially treated nanoparticles could allow to detect a single E. coli bacterium in a ground beef sample.

The potential benefits for our environment range from resource-efficient technologies reducing waste to new ways to transform and detoxify a wide variety of environmental contaminants, such as chlorinated organic solvents, organochlorine pesticides, and PCBs.

Manufactured Nanoparticles

The term "*manufactured nanoparticle*" is used here to refer to particles that have a physical size of less than 100 nm in at least two dimensions and that are deliberately produced rather than merely emerging as a by-product in activities not targeted for the production of these particles such as combustion processes or welding. While a more rigorous definition would have to take more parameters such as size distribution, diffusion diameter into account, the term nanoparticles shall be used in a rather broad sense here, including agglomerates and aggregates of the primary particles. Nanoparticles exist in various forms such as powders, suspensions or dispersed in a matrix.

In theory manufactured nanoparticles can be produced from nearly any chemical; however, most nanoparticles that are currently in use have been made from transition metals, silicon, carbon (carbon black, carbon nanotubes; fullerenes), and metal oxides. Quite a few of these nanoparticles have been produced for several decades on an industrial scale, but various new materials such as carbon nanotubes, fullerenes or quantum dots that have only been discovered within the last two decades. The development of nanomaterials is rapidly progressing and the variety of "*makes*" is increasing with considerable speed.

Nanoparticles and Human Health

The economic growth in the field of nanotechnologies will lead to an increased variety and increased volumes of engineered nanoparticles that are produced. Even if exposure assessments and data are still lacking it is foreseeable that some degree of exposure to engineered nanoparticles—for various segments of the population and for the environment—will occur to an increasingly extend over the coming years.

Keeping in mind that these "*free nanoparticles*" can enter the human body over various pathways (inhalation, ingestion or via the skin) or disperse into the environment, it is important to understand the implications for human health and the ecosystems.

To assess the risks of nanoparticles, established methods of chemical safety assessments have to be modified to address the special characteristics of nanoparticles. The main difference to the assessment of bulk materials is the fact that additional parameters like size, shape or surface properties will come into play. The same reason that makes nanoparticles technologically interesting leads to the fact that they represent a new category of (potentially) toxic substances. The interaction with the human body and their health effects are expected be different from molecules as well as from bulk materials of the same composition.

It is necessary to understand both, the hazards associated with nanomaterials and the levels of exposure, that are likely to occur. In both areas, the existing knowledge is quite limited and it will be necessary to generate and establish new data in the future.

In the following description the current status of the discussion on hazards and exposure of nanoparticles is summarized as a basis for the subsequent discussion of potential implications for the Allianz Group.

Hazards from engineered nanoparticles

When bulk materials are made into nanoparticles, they tend to become chemically more reactive – this is why they are very interesting as catalysts. Even chemically inert materials like gold or platinum are able to catalyze chemical reactions in nano-powder form.

Many studies indicate that nanoparticles generally are more toxic when incorporated into the human body than larger particles of the same materials. Experts are overwhelmingly of the opinion that the adverse effects of nanoparticles cannot be reliably predicted or derived from the known toxicity of the bulk material.

The biggest concern is that free nanoparticles or nanotubes could be inhaled, absorbed through the skin or ingested.

Uptake of nanoparticles via the lung

Most nano-sized spherical solid materials can easily enter the lungs and reach the alveoli. Inhaled particles can have two major effects on the human body:

1. Their primary toxic effect is to induce inflammation in the respiratory tract, causing tissue damage and subsequent systemic effects. The property that drives the inflammogenicity of nanoparticles is unknown but is expected relate to particle surface area and number of particles. Nanoparticles can impair the ability of macrophages to phagocytose and clear particles, and this may contribute to inflammatory reactions,
2. Transport through the blood stream to other vital organs or tissues of the body. This may result in cardiovascular and other extra-pulmonary effects.

Some scientists have compared nanotubes with asbestos in terms of risks, because they resemble asbestos fiber in their needle like shape. The concern is particularly applicable to fibers of high biopersistence. Although the comparison seems plausible, it has been pointed out, that the nanotube fibers tend to clump together rather than exist as single fibers, thus possibly significantly reducing exposure and their potential to do harm.

Uptake via other routes

The scientific literature shows that particles in the nanosize range can also enter the human body via other pathways, i.e. the upper nose and the intestines.

Penetration via the skin seems less evident, but research to clarify this is under way. The chances of penetration again depend on the size and surface properties of the particles and also strongly on the point of contact. If nanoparticles penetrate the skin they might facilitate the production of reactive molecules that could lead to cell damage. There is some evidence to show that nanoparticles of titanium dioxide (used in some sun protection products) do not penetrate the skin but it is not clear whether the same conclusion holds for individuals whose skin has been damaged by sun or by common diseases such as eczema. There is insufficient information about whether other nanoparticles used in cosmetics (such as zinc oxide) penetrate the skin and there is a need for more research into this. Much of the information relating to the safety of these ingredients has been carried out by industry and is not published in the open scientific literature.

Distribution in the body

Once in the body, the distribution of the particles in the body is strongly dependent on the nanoparticle in question, e.g. the composition, the size and of the surface characteristics of the particles. It seems that translocation occurs from all uptake routes. Such translocation is facilitated by the propensity to enter cells, to cross cell membranes and to move along exons and dendrites that connect neurons. Surface coatings will have a major effect. The mobility of different types of nanoparticles requires detailed investigations, which have not yet been performed.

To give one example: it is not clear, whether nanoparticles can pass from a pregnant woman's body via the placenta into the unborn child.

It is possible that durable, biopersistent nanoparticles may accumulate in the body, in particular in the lungs, in the brain and in the liver.

For the majority of nanoparticles the toxicological, ecotoxicological data needed to perform a hazard analysis are still lacking. Even if the details are not yet clear, it is evident that the interaction with the human body will depend on various parameters such as chemical composition, particle size, surface area, biopersistence and surface coatings among others. Therefore, until a theory of the impact of nanoparticles on human health has been established, each nanomaterial should be treated individually when health hazards are evaluated. A systematic risk screening will be helpful to establish the basic knowhow to understand the interaction with the human body and the environment and to establish the theoretical framework needed.

Besides toxic effects, the interaction of nanoparticles that have entered cells opens a wide field of potential effects resulting from the interaction with cell structures such as ribosomes and DNA.

Exposure to engineered nanoparticles

Injury can be caused by chemicals only if they reach sensitive parts of a person at a sufficiently high concentration and for a sufficient length of time. Besides the physicochemical properties of the nanoparticles, the nature of exposure circumstances and the health state of the persons at risk have to be considered.

Nanoparticles exist in nature (e.g. from volcanic eruptions or forest fires) or can be produced by human activities. Intentional nanoparticles are manufactured under (normally strict) control while unintentional ones can come from high temperature combustion, explosions, mechanical abrasion or other industrial processes. The main source of unintentional nanoparticles (in this context also called ultrafine particles) is automobile traffic.

We live surrounded by nanoparticles. To provide an example: a normal room can contain 10,000 to 20,000 nanoparticles per cm^3, whilst these figures can reach 50,000 nanoparticles per cm^3 in a forest and 100,000 nanoparticles per cm^3 in urban streets.

At present, the nanoparticles originating from dedicated industrial production are marginal in relation to those produced and released unintentionally, such as through combustion processes.

However, higher levels of exposure are expected with the industrial processes in which nanoparticles are intentionally produced or used. Furthermore, as manufactured nanomaterials become more widespread in use, the range of scenarios in which exposure becomes possible will increase.

Exposure assessments for engineered nanoparticles should be able to cover all identified uses for the entire life-cycle of the individual nanoparticle. This could include:

1. Processes involved in the production;
2. Processes involved in the identified use;
3. Activities of workers related to the processes and the duration and frequency of their exposure;
4. Risk management measures to reduce or avoid exposure of humans;
5. Waste management measures; and
6. Activities of consumers and the duration and frequency of their exposure.

Nanoparticles are currently being produced in low volumes and, aside from their use in cosmetics, involve as yet little or no exposure to populations outside the workplace. However, keeping in mind the predicted growth in production volumes and the expected expansion of the product range, this situation is subject to change over the next few years.

With an increasing number of companies involved in nanotechnologies, the quality of risk management encountered in practice will be ranging from highly sophisticated to poor. The level of risk management often depends on the branch of industry, but mainly on individual management practises and management attention in the companies. Accordingly, the assessment of possible exposures must take into account poor risk management practises, especially in the absence of specific regulations.

With respect to hazards, there is enough evidence to suggest that exposure to nanoparticles, particularly to those insoluble in water, should be minimized as a precaution.

Occupational hazards

The US national nanotechnology initiative has estimated that around 20,000 researchers are working in the field of nanotechnology. For the UK, the Institute of Occupational Medicine has estimated that approximately 2,000 people are employed in new nanotechnology companies and universities where they may be potentially exposed to nanoparticles.

The primary production of nanoparticles takes places with the help of chemical and engineering processes that are well established. These are considered by various organizations to be relatively safe, except for accidental releases via leakages. During normal production, subsequent steps like product recovery and powder handling may result in respirable concentrations of *agglomerated nanoparticles*, to dermal exposure or to ingestion (mainly via hand-to-mouth contact).

So far only one study has addressed the exposure of workers to *engineered nanoparticles*. In the study into potential airborne and dermal exposure to carbon nanotubes samples were taken during typical activities in the production process like material removal, filling, pouring and clean up (techniques: laser ablation and high pressure carbon monoxide). The study has shown that the material only becomes airborne with a sufficient level of agitation. So-called *Van der Waals forces* make the nanoparticles "*sticky*" and imparts a strong tendency to form agglomerates. Airborne concentrations found while handling unrefined material were considered to be "*very low*" (between 0.7 and 53 $\mu g/m^3$). The released material seems to form larger agglomerates in the size range of 1 μm rather than leading to a high number concentration of fine particles. Up to 6 mg of nanotubes were found on individual gloves. While the gloves can minimize dermal exposure, airborne clumps of material can lead to exposure of less well protected parts of the skin.

Contrary to intuition, air filtration systems such as respiratory protective equipment should be effective when used correctly. Below a particle size of 100 nm filtration efficiency for *High Efficiency Particulate Arrestor* (HEPA) filters even increases with decreasing particle size. The reason for this is strong Brownian motion that leads to increased probability to contact the filter elements. Once a filter element is hit, strong Van der Waals forces keep the nanoparticle stuck to the surface. It has been inferred by the UK Institute of Occupational Medicine that the primary route of exposure for many nanoparticles could be dermal exposure and subsequent ingestion exposure.

Generally accepted, realistic methods for exposure assessments are still lacking for workplaces. These methods must be biologically relevant—that is, they should be able to measure the most appropriate metric that characterizes the exposure. That can be surface area in the case of airborne particles, but also mass or number of particles in the case of dermal exposure or ingestion. Size selective sampling methods will be needed to ensure that only the relevant size range is sampled. It remains a technical challenge to develop effective methods and standards for controls. For some nanoparticles it may be necessary to measure and control to very low levels on the order of ng/m^3.

As with hazard aspects of nanoparticles, also in the area of exposure there are many unknowns today and also technical challenges lie ahead. As an intermediate state between bulk material and individual molecules the characteristics of nanoparticles add some complexity to the exposure assessment (such as surface area, agglomeration, size distributions) which can be reduced once a better understanding, better measurement methods and standards for these particles will exist. It seems that existing precautionary measures are also effective for the control of nanoparticles, however in the absence of more evidence this is difficult to demonstrate.

Exposure of consumers to products

There are a number of ways in which engineered nanoparticles from products can come into direct contact with consumers during their use. In some applications these nanoparticles are active parts of the product, in some applications only an accidental exposure is possible. Due to the fact that nanoparticles in powder form are only used as intermediate products during stages of the production process, it seems likely that the exposure from consumer products will mainly occur via ingestion and via the skin. It is unlikely that engineered nanoparticles that are bound in a matrix or somehow fixed in a product are released. Manipulations like grinding or cutting do not necessarily release nanoparticles, but rather more likely particles of larger size in which the nanoparticles are still bound.

Open points linked with the exposure to nanoparticles include:

1. Which parameters characterizing the particles should be measured?
2. The development of efficient measurement methods for these parameters;
3. Testing methods for various applications and situations during use of the products;
4. Fate of various types of nanoparticles in the environment;
5. Fate of the nanoparticles when a product is destroyed, burnt or discarded; and
6. Extent of uptake in the human body.

Safety of engineered nanoparticles

With the production of engineered nanoparticles we are confronted with a new class of materials that have novel properties compared to bulk material. Information describing the health risk of engineered nanoparticles is only evolving and many questions are still open.

The uncertainties involved have to be seen against the background of the ever more demanding public views regarding the safety of products from new technologies and an increasing potential for exposure as the quantity and types of nanoparticles used in society grow.

From animal experiments and analogies to studies on incidentally produced *ultra fine particles* (such as from burning of fuels) it is possible that at least some nanoparticles are hazardous for the human body and that the exposure to these nanoparticles should be avoided or at least minimized.

One promising path to prevent potential health hazards proposed by a number of scientists is to make the particles *biodegradable*. Particles that are degradable either by water or by lysis with enzymes will greatly reduce the risks involved because they do not persist in the body. It is self-evident that the slower the particles are cleared (*high persistence*), the higher the tissue burden can be. With a short *bio-durability*, long-term effects can be minimized or even excluded. Biocompatability can serve as one major engineering parameter for nanomaterials in the future.

There is a broad variety of nanoparticles being investigated and a high number of parameters that influence the functionality of these nanoparticles as well as their interaction with the human body. It is expected that it will take several years until some major critical risk assessment issues regarding hazards and exposure can be answered. Accordingly, in a workshop in January 2005, the European Commission has identified a variety of research needs related to nanoparticles. The recommendations for research include:

1. Ecotoxicity of nanoparticles for cases of inadvented contact by children, adults and susceptible individuals;
2. Biomonitoring studies of specific biological impact of nanoparticles, according to their most likely routes of exposure;
3. Standards development and (certified) reference materials;
4. Sensors for detecting nanoparticles and assessing exposure, both stationary and portable with a focus on low-cost, highly specific sensors giving a real time response to environment an health relevant properties; and
5. A strategic concept for a comprehensive risk assessment.

It will be a challenge for industry, legislators and risk assessors to fill all relevant knowledge gaps and to construct a set of high throughput and low cost tests for nanoparticles. In that way a risk assessment of each nanoparticle should be established before the larger quantities are manufactured. This process is unlikely to be fast enough without active steering and support by governments.

The uncertainties involved—especially long term—will have to be addressed by all organizations involved in the process of the introduction of nanotechnologies. It will be a necessary for the various industries involved to perform life-cycle assessments and demonstrate the safety aspects to a broader public. Successful communication will depend on a general feeling of trust towards nanotechnologies. This in turn needs an open dialogue and interaction involving all different stakeholders, including a high degree of transparency regarding scientific results.

Nanoparticles and the Environment

As nanotechnologies move into large-scale production in many industries, it is a just a matter of time before gradual as well as accidental releases of engineered nanoparticles into the environment occur. The possible routes for an exposure of the environment range over the whole life-cycle of products and applications that contain engineered nanoparticles:

1. Discharge / leakage during production / transport and storage of intermediate and finished products;
2. Discharge / leakage from waste;
3. Release of particles during use of the products; and
4. Diffusion, transport and transformation in air, soil and water.

Some applications like cosmetic products or food ingredients will be diffuse sources of nanoparticles. In addition, certain applications such as environmental remediation with the help of nanoparticles could lead to the deliberate introduction of nanoparticles into the environment. This is an area which will probably lead to the most significant releases in terms of quantity of nanoparticles in the coming years.

The main criteria used to assess the risks of chemicals for the environment and indirectly for human health are *toxicity*, *persistence* and *bioaccumulation*. Substances that can cause direct damage to organisms (*high toxicity*), that decay very slowly in the environment (*high persistence*) and that can concentrate in fatty tissues (*high potential for bioaccumulation*) are of particular concern. For engineered nanoparticles the particular characteristics of nanomaterials will have to be taken into account for a specific risk assessment. The existing information about properties of the bulk material will not be sufficient to classify the environmental risk of the same material in the form of nanoparticles. The possible environmental effect will therefore have to be assessed specifically for each type / class of nanomaterial. Only few studies on this very complex subject exist. From a scientific point of view, the results should be seen as indications rather than a sound basis for decision making.

In the first study on the toxic effects of manufactured nanoparticles on aquatic organisms, fish (largemouth bass) were exposed to uncoated fullerene carbon 60 (C_{60}) nanoparticles. The fullerenes are one type of manufactured nanoparticle that is being produced by tons each year. Significant lipid peroxidation (*oxidation of fats*) was found in the brain of the animals after exposure to 0.5 ppm uncoated nC_{60}. The study demonstrates that manufactured nanomaterials can have adverse effects on aquatic and possibly other organisms.

Nanoscale Iron Particles

Nanoscale iron particles have been investigated as a new generation of environmental remediation technology. Due to their high surface reactivity and large surface area they can be used to transform and detoxify environmental contaminants like PCBs. Field tests in the US have shown that the nanoparticles remain reactive in soil and water for several weeks and that they can travel in groundwater as far as 20 meters. The Royal Society has called for the prohibition of the use of free nanoparticles in environmental applications until appropriate research has been undertaken.

Cadmium

A very specific environmental issue in the case of nanoparticles is their propensity to bind with other substances, possibly toxins in the environment such as Cadmium. Their high surface area can lead to adsorption of molecular contaminants. Colloids (*natural micro-* and *nanoparticles*) are known for their transport and holdings capacity of pollutants. The adsorbed pollutants could possibly be transported over longer distances / periods of time by nanoparticles.

On the other hand nanoparticles are less mobile than we intuitively might think. It seems that their movement is very case specific and that are generally less mobile than larger particles. Here again their large surface area and their maximized chemical interaction comes into play. Their sticky nature considerably slows their transport through porous media like soil.

Explosion Hazards of Nanoparticles

For many industries, the explosion of dust clouds is a potential hazard in the production process. A *dust explosion* occurs when a combustible material is dispersed in the air, forming a flammable cloud which is hit by a flame. The concentrations needed for a dust explosion are rarely seen outside of process vessels, hence most severe dust explosions start within a piece of equipment (such as mills, mixers, filters, silos).

Various materials that are not stable oxides can be involved in dust explosions, e.g. *natural organic materials* (grain, sugar, etc); *synthetic organic materials* (organic pigments, pesticides, etc) *coal and peat metals* (aluminium, zinc, iron, etc). There is a clear dependence on size and surface area of dust particles, it does however not vary linearly with the explosibility of the powders. Dust with a particle size above 0.4 mm is normally not explosive.

In certain applications, the stronger reaction of smaller particles can be used to pack more explosive power in a given volume. Today, *microsize aluminium particles* already gets incorporated into rocket fuels and bombs. *Nanoaluminium* is being investigated as advanced technology in these military applications. *Aluminium nanopowders* could also find their way into airbag gas generators. The toxicity is being investigated and has to be compared to the properties of existing propellants and their combustion products (particles and gases).

The explosibility of nanopowders has so far not received much attention in the public debate on health and safety risks of nanotechnologies.

A review of literature available on this topic has been performed by the Health and Safety Laboratory in the United Kingdom. The review could not find data for nanopowders with particle sizes in the range of 1 to 100 nm. It states that the extrapolation of existing data for larger particles to the nanosize range is not possible because of the changed physical and chemical characteristics on the nanoscale. The explosion characteristics of a representative range of nanopowders is determined since an increasing range of materials that are capable of producing explosive dust clouds are being produced as nanopowders. That large quantities of combustible nanoparticles do not become airborne until the explosion hazard has been properly evaluated. As long as nanopowders are produced in small quantities of grams, the explosion hazard will be negligible.

Besides property damage, workers safety and business continuity issues, one major concern about the explosion of nanopowders is the release of larger quantities of nanomaterial into the environment and the resultant potential pollution problem.

Self Replication of Miniature Machines

The Coming Era of Nanotechnology," in which Eric Drexler in 1986 imagined the fabrication of molecular machines. These machines would be able to produce any (macroscopic) item from molecular building blocks. For this plan to work, these machines would have to be able to produce machines of their own kind, a process called *self replication*. His most compelling argument about the feasibility of these machines is the observation that biology gives us many examples of nanoscale machines that function on this scale.

The idea that engineering a synthetic form of life with self-replicating machines has, in turn, created the fear that once designed, these nano-robots could spread across the biosphere. Drexler called this scenario the "*gray goo*".

This view has been challenged by many scientists. Most scientists have dismissed the "*gray goo*" scenario as "*science fiction*". There are fundamental questions that have not been resolved so far, like the energy management, the strong surface forces on the nanoscale or Brownian motion. After all, biological systems in our environment have been optimized over billions of years of evolution and make extensive use of the particular physical characteristics that govern the nanoscale world, which we are just beginning to understand. Our knowledge even of the processes in a simple cell is limited.

It seems therefore safe to say that the construction of self-replicating nano-robots will remain beyond our capabilities for the foreseeable future. From an insurance perspective the risks associated with self-replication of machines will therefore remain futuristic and will—in all probability—have no relevance over the next decade.

Regulatory Considerations

It is thought that nanotechnology exploits properties not generally seen in large-scale solids of the same chemical composition, and that these same properties have led to health and safety concerns. For example, the high surface reactivity of nanoparticles and the ability to cross cell membranes might have negative human health impacts. On the other hand, not all nanoparticles, and not all uses of nanotechnology, will necessarily lead to new human health or environmental hazards. A distinction can be already be made, for example, between free and fixed nanoparticles. Fixed nanoparticles are less likely to pose a problem because they are immobilized within a matrix and cannot freely move or disperse within the human body or the environment. In other words, there is likely to be low human exposure.

It may be possible to assess many products involving nanoparticles from the chemicals industry using existing mechanisms or techniques for risk assessment, but it may often be necessary to adapt existing techniques or devise new risk assessment methods. In any case, it will be important to address safety issues in a proactive way in order to avoid public uneasiness about the new technology as a result of a lack of information and attention to safety aspects which could lead to limitations on the safe development of a promising technology.

Drug delivery, which may turn out to be the most important area economically, will take some time to come to market because of the time needed to complete approval processes. The drug example raises the possibility that nanotechnology may suffer from a new problem, termed "*toxicology bottleneck*" as the science advances too fast for the risk assessment processes of organizations. This is further complicated by the difficulty in both measuring and modelling nanoscale particle behaviours in both air and fluids.

The effects of inhaling free manufactured nanoparticles have not been studied extensively. Analogies with results from studies on exposure to other small particles such as the pollutant nanoparticles in urban air and mineral dusts in some workplaces suggest that at least some manufactured nanoparticles will be more toxic per unit of mass than larger particles of the same chemical. It also seems likely that nanoparticles will penetrate cells more readily than larger particles. The report also emphasizes the diversity of technologies lumped together under the term 'nanotechnology' and the implications of that diversity for the approach to public dialogue, research and regulation.

Planned steps to ensure a safe and ethical development of nanotechnologies include:

1. Setting up a research coordination group to investigate risks from nanoparticles.
2. Initiatives for public dialogue to help the scientific community and the public to explore issues relating to the regulation of nanotechnologies.

It is also planned to review the adequacy of the current regulatory frameworks.

1. The assessment of risks associated with medicines and medical devices.
2. The safety of unbound nanoparticles in cosmetics and other consumer products.
3. Disclosure of testing methodologies used by industry.

4. Labelling requirements on consumer products.
5. Sector specific regulations for products of nanotechnologies in addition to REACH at a European level.

At present no specific regulations exist which refer specifically to the production and use of nanoparticles either for workers or consumers' safety or for environmental protection. Current rules and operational practices are applied.

There are several research projects funded by the European Commission which deal with nanoparticles (eco)toxicity and risk. The *NanoDerm* project investigates the quality of skin as a barrier to ultrafine particles. *NanoSafe* and *NanoSafe2* (in negotiation) study risk in production and use of nanoparticles. *NanoPathology* investigates the role of micro- and nanoparticles in inducing biomaterial disease. Other projects are currently being negotiated such as the specific support action *NanoTox* that aims to provide support for the elucidation of the toxicological impact of nanoparticles on human health and the environment, and the coordinating action Impart that aims to improve understanding of the impact of nanoparticles on human health and the environment. Moreover, the German project *NanoCare* will develop an inhalation toxicity model for testing of nanomaterials.

Industrial Position

The industrial working group aims at the successful realization of the economical and technological chances by initiation of suitable measures, taking into account ethical, ecological, social and health care aspects. The tasks of the working group are:

1. To identify and prioritize research topics which have to be addressed in order to assess possible risks of nanomaterials.
2. To prepare project proposals on the basis of identified research topics and to support their realization.
3. International cooperation to gain synergies and to widen the database.
4. Dialogue with stakeholders.
5. Promoting related communication.

On the risk management agenda, the emphasis of the working group lies on identifying possible risks of chemical nanotechnology focusing on nanoparticles and nanotubes in both their free and bounded state. The working group favours a structured approach by assessing hazards and exposure and implementing risk management measures such as:

1. Occupational protection measures for workers.

2. Use of embedded nanomaterials, use of nanomaterials chemical fixed / covered at surfaces for consumer products as well as individual toxicological testing and governmental authorization.

This approach reflects the attitude of industry representatives, who generally call to explore the potential risks case by case, application by application and material by material. One basis for this approach is establishing databases for the scientific evaluation of risks. It is not clear whether public access to this kind of databases is intended.

5

Diamond Teeth

Recent, exciting strides in permanently attaching false teeth to the jaw, or using "*space-age*" plastic to restore an old tooth to its natural strength and beauty, can make it easy to forget that not too long ago, millions of people in the United States had all their teeth out by the age of twenty-five or thirty in order to put an end to bothersome, rotting teeth. In fact, through the 1950s, families that could afford a dowry would often provide for the purchase of a spouse's dentures. Today it seems that everyone over the age of fifty is anxious, to a degree, about the prospect of losing some or all of their teeth.

Is tooth loss inevitable? No. When nanotechnology arrives—two or three decades hence—it will arrest the genetics behind tooth loss and the accompanying thinning and deterioration of the jaw bone. Moreover, nanotechnology will be able to reverse this process of degradation. Let's look at some of the latest advances in the field of dentistry and then consider how a visit to the dentist will be irrevocably altered when nanotechnology becomes a reality.

Tooth Replacement

Dental implants, or *permanently attached false teeth*, have become very popular. Metal anchoring units surgically implanted into a patients' jaw allow the bone to flow around and bond to them. This bonding process, known as *osseointegration,* can take up to six months. Once complete, a dentist attaches one or more artificial teeth to these strong anchors. After several visits to the dentist over a period of several months, the satisfied patient with a brand-new smile is able to chew on an apple the way he or she did when young and, well, full of teeth.

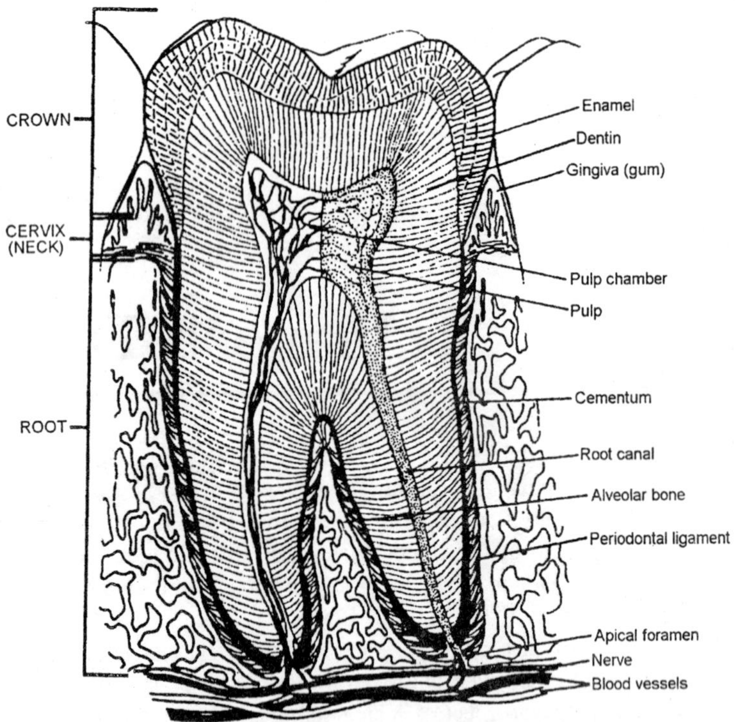

Fig. 5.1. Structure of tooth.

Unfortunately, malpractice suits related to this practice have increased, partly because the implant process is sometimes carried out too swiftly or the implants are improperly fitted. In some cases, the procedure should never have been recommended. With aging, a patient's jaws can become so weak or diseased that implants cannot find a firm foundation; implant technology available today does not replace the jaw bone or mitigate bone thinning as we get older.

Dentists have recently begun to use computer-aided design and manufacturing (CAD-CAM) systems to fabricate porcelain caps. With one end of a fiber-optic light wand plugged into a desktop computer, the dentist moves the free end over the tooth. Instantaneous measurements of the tooth contours (before and after the tooth is drilled) are transmitted via the computer to a portable milling machine, where a tooth-coloured cap is fabricated. It is then permanently cemented onto the prepared tooth—all within 90 minutes. No more suffering for two weeks with an ill-fitting, temporary tooth covering all the while dreading another trip to the dentist for the final procedure.

A genetic approach, only now on the "*biological drawing board*," could make even this CAD-DAM procedure obsolete. By the turn of the century, your friendly neighbourhood dentist will routinely repair that cracked molar, not with computer-designed caps, but simply by adding a tooth-coloured "*paste*"—derived from cloned enamel genes—to almost any broken-down tooth. Thanks to a few small alterations in the DNA, this biocompatible material could be stronger than the original enamel and more resistant to cavities.

Dental Care in Nanofuture

Let's get into a dental time-machine set for the year 2020 and visit that posh dental suite of Dr. Harvey Smile-Maker. His office is located, naturally enough, in Southern California, where so many new dental techniques are firs attempted–for better or worse.

Dr. Smile-Maker peers into the oral cavity of his new patient. Mr. John Garbage-Mouth. Mr. Garabage-Mouth's teeth are in state of chaos. He steadfastly refuses to use his robotic tooth-flosser before retiring for the evening. His habit of eating chocolates has accelerated as already severe case of tooth rot. The last straw was when his horrific mouth odor finally drove his girlfriend to leave him. The man is desperate for (among other things) some good dental care.

Mr. Garbage-Mouth timidly reclines in an air-cushioned dental chair that is computerized to custom fit the contours of his body. As our patient hopefully nods at the holographic "we cater to cowards" sign, the dentist adjusts a handheld, *portable positron-emission tomographic* (PET) *machine*. The PET scan will show the precise, three-dimensional characteristics of the toothless regions of his mouth and will summarize the following: normal/abnormal bone and gum densities, all vessels, and the specific sites where further tooth or jawbone loss is most likely to occur. All this information is used in conjunction with other scanning devices that feed the data directly into a computer. Powerful "*expert system*" software almost instantly determines optimal jawbone sites and the precise amount of biological ingredients to be used for tooth rehabilitation.

The treatment begins. Patches attached to the patient's gums send out electrical signals that deaden the appropriate gum and jaw nerves within seconds (injections of anesthetic will be a thing of the past). A computerized robotic arm designed to work inside the mouth begins drilling tiny, cylindrical holes into the jawbone at positions selected by the PET scan. Then the robotic arm grasps a syringe containing a few drops of "seed" material containing uncountable programmed assembler molecules and injects them into the drilled sites. A dental impression tray containing the construction materials necessary for building the tooth is carefully fitted over the site.

This fantastic technology resembles a disposable plastic dental-impression tray of the 1990s that was used to make a mold of the prepared tooth so that the dental lab could fabricate a cap. On closer inspection, we see the interior or backbone of the tray is lined with trillions of molecular-sized computers. Working in parallel, they precisely regulate the flow of construction materials—mixtures of artificial calcium hydroxy-apatite crystals (the molecules that bone is made of) and the natural and semisynthetic organic molecules, as well as other molecules never seen before—as they come in contact with the seed material within the jawbone sites. The process is initiated and powered by an excimer laser pulsing through a fiber-optic bundle. Thus excited, the nanoseeds configure the scaffolding for a new tooth, consuming the provided construction materials flowing from the tray.

"Bleep, burp, blam!" Before you can say "*Dental floss*," a beautiful, perfectly shaped tooth begins to take form. And don't worry about long dental appointments: "*nanodentology*" will produce pearly whites within an easy morning appointment. Or just about long enough for our patient to get those chocolate cravings again.

This wonderfully *high-tech molar* will duplicate or improve upon the clinical and morphological characteristics of the original, unworn tooth that existed prior to decay and breakdown. Unlike the cloned-gel material discussed earlier, there will never be the danger of tooth fracture or decay at the interface of natural and artificial tooth substance. The entire replacement tooth, root and crowns, will consist of a single, solid, cavity-proof matrix.

Constructed with atomic precision, the new tooth will be a substantial improvement—considering strength and durability—over mother nature. Biting an apple will feel as comfortable and satisfying as if you had the teeth of a teenager. What a far cry this dental era will be from the 1990s when it took stressful surgery, discomfort, and a wait of up to nine months for an implanted artificial tooth to achieve a similarly strong bite.

Diamond Jaws

Nanotechnology will deliver the holy grail of dentistry: long-lasting, cavity-free teeth. But just a few months or years later, advanced nanotechnology will deliver another coup: arresting or neutralizing the genetics behind a degenerating, aging jawline. For if we are able to fabricate perfect, ageless teeth, could not the same technique be used to augment and strengthen a thinning mandible? And what material would be the strongest and most durable to use for this procedure? *Diamond.*

Our patient could be an eighty-year-old with few or not teeth and a jaw that the ravages of time had narrowed to only the thickness of a pencil. In this procedure, uncountable assembler molecules from the dental tray are quickly depositing a diamond substructure that fills in and augments the disease-deficient and thinning areas of the patient's jaw. At the conclusion of this treatment, the patient could well end up with the youthful jaws of a twenty-year-old—or better. Forever freed from the ravages of a deteriorating jawbone, not to mention bone cancers, numerous bone disorders and other distasteful age-related phenomena, we could enjoy the jaws, teeth, and smile of youth, perhaps for decades.

We could eventually see the replacement of the entire jaw and teeth with a *diamondoid matrix* but why stop there? We can expand this approach to improve or replace the body's entire skeletal structure. We've seen *Batman*, *Superman*, even *Lawnmower Man*—look out now for *Diamond Man*. Only this amazing human will be each of us, an Everyman, finally freed from bad teeth, worn-out jaws, and other bone maladies.

What an era the human race is fast approaching! A time when crutches, bone casts, and dentures will exist only in museums. And if you decided to never brush or floss again, the worst you might get is a case of a diamond breath. To help you hold out for the nanotech miracles of your local dentist in a couple decades hence, floss and brush well. The world of dentistry is about to offer exciting improvements—all for a healthier you.

6

Artificial Intelligence

Nonmobile computers are already more plentiful than robots and will always be cheaper for the same processing power, so stationary computers as smart as humans will probably arrive a bit sooner than human-level robots. What is more, stationary computers can do desk work and deal in the business or technical world without needing all the sensory and motor mechanisms that humans have for physical activities. So perhaps they will be a little easier to build, and get here just that little bit earlier.

Artificial intelligence techniques already pervade the business world (although often they are called other things). Every time you make a long-distance phone call or use a credit card, some computer program is analyzing the transaction and deciding whether your card is being used fraudulently. Some of these are simple, dumb, rule-of-thumb programs—but some are quite subtle, and better than humans at picking fraudulent from legitimate usage based on the customer's usage patterns and typical fraud usage patterns. And what is more, the systems do this for all the literally hundreds of millions of calls and card transactions that happen every day, in real time. Think of it: some of the largest companies in the world base a substantial bulk of their sell/don't sell decisions on the word of a machine, and one following not simple rules (in the cases that work well), but sophisticated statistics-based inference procedures whose math the businessmen don't begin to understand.

When machines can do that well at other kinds of decisions, they will be put to use. Suppose similar techniques worked for hire/don't hire decisions. Or fire/don't fire decisions.

CYBERNETICS

In the 1700s, at the dawn of the industrial revolution, before steam power became the prime mover of machinery, the major sources of power for stationary mills were wind and water. Water power was straight forward to harness and generally even and reliable; but many places don't have an appropriate stream. Wind power is available many places water is not, but there are two problems: the wind does not always blow at the same speed, and it does not always blow in the same direction.

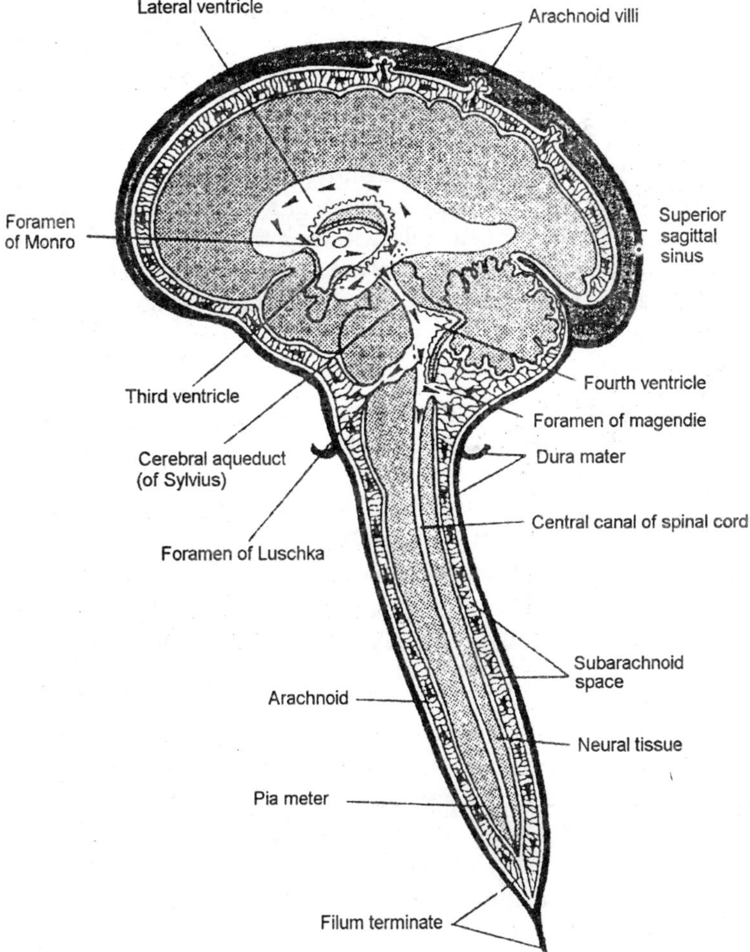

Fig. 6.1. Human brain.

Windmills were often built with some ingenious additions to the main spindle and sails. In 1745 Edmund Lee patented the automatic fantail, a small secondary fan, set at right angles to the big one. The small fan is connected to gears that turn the whole windmill on its base. The trick is that the small fan is edge—on to the wind, and thus does not do any turning, except when the big fan is facing squarely into the wind. The small fan thus supplied not only the power, but the smarts, to keep the windmill properly oriented. Windmills also had automatic reefing arrangements, in which a system of weights holds the sails to the wind, but if the mill spins too fast, the weights are pulled up by centrifugal force and the sails are partially furled. Sources disagree on the origins of these, but they were in widespread use in the eighteenth century.

These devices appear to have been the origin, or at least the gateway into modern Western technology, of mechanical regulation by feedback. Feedback has become something of a buzzword in organization and communication, but before that, it was a buzzword in engineering. In the engineering sense, it simply means a process in which the output of a machine is measured, and the measurement used to control the machine.

Thermostat

One of the simplest and most common examples of feedback in control is the *thermostat*. A typical home heating thermostat is a combination of a thermometer and a switch that turns the furnace on or off. As the temperature in the house rises above the desired temperature, called the set point, the thermostat turns the furnace off; as it falls below the set point, it turns the furnace on.

There is some lag in this system, producing what's called *hysteresis*. First, the thermostat itself does not actually turn the furnace on and off at the same temperature. For example, it may turn it off as the temperature rises above 72, and only turn it on again as it falls below 70. In addition, it may take some time for the heat to seep out from the radiators across the room to the thermostat. As a result of both of these, the furnace does not just flicker on and off but stays on or off for longer periods of time, which is good for the furnace but causes the temperature to swing up and down.

Hold an arm out at full length, and stick a finger out sideways so you can see it. Now watch the finger with reference to some fixed point in the background behind it, and you'll find that it is not actually rock steady. It wanders around the point you are trying to hold it, in a way very similar to, and for the same reason as, the temperature fluctuates around the set point of the thermostat. In other words, your own control of your muscles uses feedback.

Let's assume for the sake of argument that you would rather have the temperature (or the arm) be steady. One thing you can do is get a furnace (or circulator pump or fan) that is not simply on or off, but which can be set to run faster or slower. The problem now is that the thermostat has a harder job to do. It is not enough to turn on or off—it has to tell the furnace how bard to run. (Your nervous system already does this, of course—you don't hold an arm in position by alternately jerking your muscles absolutely tight and then going completely limp.)

It would help if we knew the speed to run the furnace that kept the house at an even temperature. Then the thermostat could have the furnace run at that speed at the set point, and run harder if it was colder, to warm up to the set point, and run slower if it were warmer, so as to cool down. Such a scheme is called *proportional control.* Note both of the windmill control schemes described above were examples of proportional control—more sophisticated than many modern-day thermostats!

The problem with proportional control is not only that you don't know in advance what that magic even-temperature speed is, but that it changes. Changing temperature outside, opening of doors or windows, and/or other heat sources inside can all throw off even a correct initial estimate. For example, if someone is left a window open, you need to run harder to maintain the same temperature. And indeed, a simple proportional control will allow the temperature to vary some what with the conditions.

A smarter system might have a thermometer outside as well as inside, to let the thermostat guess how hard it needs to run the furnace. We can imagine sensors on the windows and doors, and sensors to tell how brightly the sun is shining and whether the roof is snow covered. In reality, of course, houses don't need nearly as sophisticated a control, but some things do. Animals—and humans—are a case in point.

Typical home heating systems are not set up so that the room there most at controls the furnace; it controls a circulating pump, and there is a thermostat on the boiler that actually controls the Furnace. It would be possible, although it is not common, to have a proportional control of one level of such a duplex system control the set point of the other. It is not common in house heating, that is; in more complex systems, it is much more frequent. The nervous system of an animal, or even an insect, is a complex hierarchy of feedback loops controlling feedback loops.

It is clear that the character of nervous systems as feedback control systems was one of the prime motivators for *cybernetics*. Cybernetics was the original mathematical treatment of feedback for control in higher math, done by Norbert Wiener in the World war II time-frame. It was successfully used in the design of automatic radar-controlled antiaircraft guns, among other things. There were high hopes that the theory could be extended to form a basis for understanding everything from animals to economies. And in fact, dynamic feedback loops do pervade a wide range of phenomena. But the mathematics did not, in the end, suffice to put a broad, common base under all of them, and the term *cybernetics* and many of its grander dreams faded away.

Meanwhile, control theory, as a more restricted branch of engineering, flourished and is now considered one of the fundamentals of the field. Feedback is extensively used in mechanical, hydraulic, and electronic systems, and its analysis is not considered anything other than part of the engineer's standard toolkit. It can be used to design active creatures of considerable complexity. It is clearly present all throughout the structure of actual nervous systems. The broad range of mental phenomena such as hypnosis, trances, dreaming, and the variations of consciousness ranging from intense concentration to woolgathering speaks tellingly of an organization much more like a network of feedback loops than a straightforward algorithm.

Good Old Fashioned AI

The history of science is inseparable from the notion of reductionism. Take a complex natural phenomenon—the classic one is the courses of the stars and planets in the heavens. Reduce it to a simple model—in the case of astronomy. The model was ultimately a very few numbers for each planet, denoting its mass and position—and simple rules (often in the form of differential equations) for manipulating the models.

As *artificial intelligence* (AI) got its start, naturally an attempt was made to do the same thing that had worked so well for the physical sciences. A classic example was language translation. A very simplistic model of translating from one human language to another is to find the word in the target language that corresponds to each word of the source language text and string them together to form the output text.

As you might expect, such an approach does not work very well. The results are perhaps best characterized by the jokes that come down to us from that period: "Some researchers decided to test their translating computer. They gave it the sentence 'The spirit is willing, but the flesh is weak' to translate into Russian, and that they had it translate the result back into English. What the computer told them at the end was, 'The vodka is strong, but the meat is rotten." Some phenomena are harder to model than others, after all. Physics took a bit longer to handle electromagnetics, or fluid flow, than gravitational dynamics at the planetary scale.

So artificial intelligence researchers went on to other things; and some really impressive results began to appear. AI programs at the complexity of one Ph.D. thesis began to be able to do things like solve freshman calculus problems, play chess, or do analogy problems like those found in IQ tests.

Perhaps the crowning achievement of this period was Terry Winograd's SHRDLU program, which was capable of playing with (simulated) blocks, holding a conversation about what it was doing, and to all appearances really understanding what its correspondent was saying in simple conversational English.

SHRDLU was a conceptual breakthrough because it contained a semantic model—an inner program that could be used as a simulator in a way that gave meaning, if only the simplistic glimmerings of it, to the symbols the program was manipulating. In other words, SHRDLU could seriously be said to understand the things it talked about, if only in a very limited way.

What it could not do is learn (except by being given an explicit definition for a new word). This takes nothing away from SHRDLU—it was every it the milestone I described, and it was written in a day when all the computers at MIT together could not match one of today's PCs for raw bit—flipping power and storage capacity.

The problem is that in the third years since SHRDLU, no one has come along and finished the job. The things SHRDLU demonstarted-parsing, disambiguation based on a prefabricated semantic world model, manipulation of such a model, and so forth—have been improved almost beyond recognition. We have chess—playing machines that can best any human, but life is not chess. We have programs that do amazing searches in amounts of data much bigger than any human could handle. We have systems that are indispensable for the design of ever-more-complicated machines, and that facilitate the design of ever-more-complex software. At the experimental level, computers can converse in spoken English, recognize faces seen with video cameras, and control robot bodies that walk on two feet.

The AI community has taken a somewhat justified stance in response to its critics, that whatever it was that computers could not do was considered AI, while whatever they succeeded in reducing to practice was promptly moved to the field of "known *software technique*" and not considered AI anymore. So AI was never given credit for what it accomplished, but was always measured on what it had not done.

After all, in the 1960s, playing chess and doing symbolic algebra was considered AI. In the 1970s, image and natural language processing were considered AI; in the 1980s, inferential databases and rule-based diagnostic and configuration systems were consider AI; in the 1990s, robots that drove cars or walked were considered AI.

On the other hand, in the 1950s, *translating languages* was considered AI, and it still is. And to be fair in the other direction, AI has generally defined itself in such a way that it would naturally be measured by the difference between what computers can do and what humans can. So how far has it gotten? The things that have been achieved form an impressive list; but they are a small fraction of the things that humans can do. Let's take a look at the things that computers still cannot do.

In the field of vision, for example, it is now fairly well understood how to duplicate the preprocessing that the few layers of nerve cells in the retina do to the visual image before sending it along the optic never to the brain. This may sound trivial, but it requires one thousand times the processing power available to the early AI researches.

To continue, the reverse engineering of the visual system has proceeded to the point of being able to separate and distinguish objects in the visual field, and place them in a three dimensional world to some extent; but the current state of the art does not seem capable of general identification, that is, telling you what each object is.

The current-day descendants of SHRDLU can converse in colloquial English about specific, well define dl topics, and current-day expert systems do professional quality jobs on some impressively complex and subtle tasks. But if you get these systems out of their field of expertise, they don't have the ability to learn the new sniff or even react with common sense and general intelligence to what they experience.

Robotics

In the field of *robotics*, the state of the art ranges from assembly manipulators in factories, which do simple manipulations at blinding speeds, to Asimo, the famous Honda robot in the shape of a human, which walks, albeit slowly. In between are the autonomous soccer teams (of small, wheeled robots) that play respectable games of table-top soccer (where speed and strategy are of equal importance).

And yet, present-day robots are still very primitive by the standards of animals, or even insects. Rodney Brooks, head of MIT's AI Lab and famous for creating several of the revolutionary robots of the 1980s and 1990s, goes so far as to say that we are still missing some basic key concept, in the way we were missing calculus while attempting to understand the motions of the planets, before Newton and Leibniz filled us in on it.

(It is just a little bit ironic that Brooks made his name, and to some extent revolutionized robotics, by tossing out the mechanisms that AI had always assumed needed to he there, and building robots that ran on reflexes rather than logic.)

However, it does not seem to he the higher reflective functions that Brooks is worried about; in fact, when he characterizes his notion of what is lacking, he holds open the possibility that we may just need more parameter twiddling, or more computer power, or any other normal, expected progress in the field. Still, he seems to think that what is needed is that a light bulb should go off in some researcher's head and a "Eureka!" moment would then occur.

When you watch the actual robots or interact with the programs, it is not easy to decide which side of this question to agree with. On the one hand, they are enormously, almost incredibly, better than they were a quarter century ago. And yet, they are still so awkward, "lumpen," or clueless that there seems a difference in kind, not degree, between them and the natural systems we are trying to emulate.

Perhaps one way to characterize what the difference feels like is that it is like comparing a pencil line drawing to a colour photograph. The drawing can be enormously suggestive of the picture, clearly revealing a depth of perception and skill on the part of the artist, as the AI programs do about their creators. Yet there are a thousand subleties the drawing does not capture that are evident in the picture.

Part of this is clearly due to the fact that AI systems do not have the computational power to throw around that living ones do; brains still outperform silicon, ounce for ounce and watt for watt, by a few orders of magnitude. But this is, and has been, changing, and it is very likely that particular constraint will be gone inside two decades.

Another reason for the parsimony of artificial systems is more endemic to their nature; it is the same reason the pencil drawing is more parsimonious than the photo. The drawing was done one mark at a time by a person; each detail, and its relation to each other detail, occupied his consciousness at some point. The more detailed the drawing, the more effort was required in its construction. The same is true of AI programs. (Yes, a multiperson development team, as is used for any significant commercial software, does help. This is still not hugely prevalent in AI so far, but as AI moves out of the academic into the commercial world it will increase.)

Artificial Intelligence in New Century

As 1999 turned to 2000, people across the world gathered to celebrate the new year, the new century, the new millennium. Once the parties were over, some wags, including your humble narrator, pointed out that there was still a year to go and that we'd have to do it all over again. Centuries begin with the "01" years because there was not a year 0, and so the first century began with year 1.

Numbers can be used in two subtly different ways. First there is counting, to tell how many of something you have. But there is also naming. If, for example, I live at 6742 Waycaster Street, it does not mean that there are 6742 houses from one end of the street to mine. The number 6742 acts like a name. And it would still act like a name if there were no gaps; the first house would be named 1, the second 2, and so forth.

Look at a ruler. You know there are twelve inches on it. Yet physically, there are either eleven marks dividing the inches, or thirteen counting the ends; not exactly twelve. The numbers are measurements denoting the points between the inches, not names of the inches (in which case, of course, there would be exactly twelve). Yet the year numbers are used as names of the years: hence the confusion.

When you are programming a computer, you have to understand exactly how you are using your numbers or you'll introduce bugs your program. (Bugs arising from this particular confusion are some times called fencepost errors.) And yet most people manage to live their lives happily without ever being aware of the distinction, much less applying it correctly in each situation. And what is more, we usually get it right anyway—we go to the right house, using the number as a name, and measure the room, using numbers as measures, and don't even notice the difference. We have a concept of counting that includes both, and somehow it seems to work.

This could not possibly work if you were built the way current soft ware is. It is built from the bottom up, and all those details have to have been noticed, and gotten right, or the program does not work at all. We humans seem to be able to come at things from the top down, making finer distinctions if necessary but getting by with no trouble on general notions in most cases. Conventional software, such as the applications on your PC, just runs in the opposite direction from the start. Literally millions of tiny details, each critical to get exactly right, combine to give the overall effect that you see.

Besides coming at the problem from the wrong direction, there is the business of the amount of detail described above. Because software is built from the bottom up, building an AI program is much more like building the brain a neuron at a time than like starting from some condensed, high-level description such as the genome. Viewed from this perspective, a mature human brain is much more complex than the high-level description—perhaps a million times more complex. Where did that complexity come from?

The best way to describe it is that the brain is autogenous: it grows in patterns that depend on both the original instructions and on the inputs from the senses. It grows, it learns, it builds itself.

In other words, the nature of the code in AI programs is different in a fundamental way from the nature of the code in the genome that builds the brain (even in the highest, most abstract, analogy): the AI is a program for doing cognitive tasks, but the genome is a program for building a machine that does cognitive tasks. Or perhaps for building a machine that builds machines to do cognitive tasks.

Nontraditional Artificial Intelligence

While symbolic AI was taking off in the 1960s, a number of biologically inspired alternatives got left by the wayside. Besides cybernetics, they included neural networks and genetic algorithms. When purely symbolic systems began bumping into ceilings in the 1980s, the other approaches enjoyed a new wave of popularity and investigation. Given considerable new scope by the processing power that began to become available by the 1990s, these methods made great strides.

The specifics of these methods are not important to this discussion; there are many of them, each with lots of variations. What they have in common is that they lack the brittleness of the rule-based approaches, but they also lack the ability to handle large complex problems. One of the features of the nontraditional approaches is often that they learn the function they are supposed to perform, an advantage, but by the same token they require masses of training data, which can be a drawback.

It often turns out, by the way, that the methods derived from biological (and physical, and other) analogies, which work, can be analyzed so as to reduce the model to something more straightforwardly statistical. For example, neural networks can be trained much more efficiently with an ODE (ordinary differential equation) solver than the biologically inspired "back-propagation" algorithm. Methods such as Bayesian occupancy grids for robot navigation simply adopt the statistical basis to begin with (though perhaps inspired by neural maps), but have much the same feel, with many of the same strengths and weaknesses.

What is needed, of course (and this has been obvious to AI practitioners for decades), is to combine the techniques in such a way as to add the flexibility and learning ability of the biologically inspired approaches to the structured complexity of the symbolic ones. Many attempts have been made to do this in various ways.

Copycat

One of the more interesting ones was the series of programs produced by Douglas Hofstadter's group at Indiana University. The most famous was Copycat, by Melanie Mitchell. Copycat was a program that answered letter-string analogy problems. (For example, if ABC becomes ABD, what does PQR become?) The goal of Copycat was to explore several techniques that differed from straightforward algorithmic, symbolic ones in favour of flexibility ("fluidity," in the group's phrasing).

The essence of what Copycat did was to build interpretive structure, like a classical symbolic AI program, but using a probabilistic method based on a biological model: the inner workings of a cell. In other words, Copycat worked internally like a soft version of diffusive transport and self assembly. Thus Copycat did have, within its microdomain, the combination of flexibility with symbolic complexity that AI needs.

Copycat, like SHRDLU, illuminates a significant building block of cognition. In SHRDLU's case it was understanding as a function of a model. In Copycat's case it is a general operation we can call analogical quadrature. Copycat's problems are like the analogy problems you often see on IQ tests. The point is to make an analogy between two things and carry the analogy through a transformation of one to transform the other appropriately. We can imagine, though no one has written so far a scaled-up industrial-strength version of Copycat that could do analogical quadrature with the same insight, flexibility, and occasional inventiveness that Copycat itself brought to letter sequences, but do it to data structures in a full-fledged modeling language able to describe objects, situations, actions, and events.

Another similarity Copycat has with SHRDLU, unfortunately; is that it does not learn its capabilities; its total stock of knowledge, carefully tuned and coded, is set up by the human author. Again, this in no way detracts from what it does do; this is just to remind you that many deep problems remain to be solved.

A Symbolic Servo

Back in the dawning days of AI, two of the founding fathers, A. Newell and Herbert Simon, created one of the first landmark programs, called GPS (meaning General Problem Solver). GPS worked by being given a description (in a high-level, symbolic language of course) of a current state and a goal, and having a list of possible actions, it would choose the one that would most reduce the difference between the state and goal, and continue, starting with the new state. The story is took that when AI founding father John McCarthy first saw GPS, he exclaimed, "Why; it's a symbolic servo!" A servo is feedback control loop as described in connection with cybernetics above. GPS works, as well as any symbolic AI system does—its descendants include Soar, the most successful general-purpose classic AI system to date. It is clear that lower levels of the nervous system act like conventional servos. It is also clear that among the other things going on in the upper reaches of cognition, something like GPS's (or Soar's) symbolic manipulation is happening. What if we could find some general mechanism that would bridge the gap? Then we would be able to think about a mental architecture that was a complex hierarchy of feedback loops, as cybernetics and neurophysiology suggest, but retain the ability to do the symbolic stuff that has been the weakness of that level of theory.

Here is a possible candidate. It is based on analogical quadrature and an "industrial-strength Copycat." It is called a *sigma servo* or sigma for short, which stands for Situation, Goal, Memory Action. Imagine it is embodied in some small chunk of the brain (which consists of many such chunks, complexly interconnected). It has as inputs sets of sensory signals, possibly connected directly to sensory organs, possibly highly interpreted, out in any case signals that may be taken to define a situation. It has nerve signals coming from above in the hierarchy, which can he thought of as a goal. It has outgoing signals it can generate that are thought of as its actions. And it has a memory of its past activity which is organized as triples of situations, actions, and the new situation that followed after performing the action, which we will call the result.

The sigma now does an analogical quadrature between its current inputs and every remembered triple. It picks the one that has the best overall match (or interpolates between multiple close misses) and does the new action. (In a somewhat more developed version there is an aversive flip side, where results are remembered as painful and corresponding actions suppressed, as well.)

The level at which the *sigma servo* operates is now completely general. It has its own memory, written in its own language. If situation and actions are simple numbers, quadrature works out to be interpolation and the servo is a proportional controller. This is typically as sophisticated as is needed for, say, controlling a single muscle. In this case, as with the higher level case in general the situation is augmented with a history gathering and remembering mechanism.

Note that at the lower levels, it can be continuous in operation. As the input signal varies over time, the output varies, too. The higher and more symbolic it gets, the more discontinuous jumps there must be, until you get to something more like a finite state machine. Note that two of today's leading roboticists, Albus and Brooks, favour cognitive architecture that are hierarchical feedback networks of finite state machines.

With highly abstract, structured, *symbolic memories*, the sigma servo becomes a processing engine of the AI sort. Indeed, the situation begins to look a lot like what Minsky called a *frame*. And variations in the kinds of signals processed and remembered allow the servo to be adapted to a wide variety of tasks. More to the point, there not just one of them, but a complex network, and the situation for some is a description of what the others are doing—and their actions consist of setting the others' goals and/or characterizing each other's results.

One last point: the triples in memory, particularly at higher levels, don't come just from one is personal experience but from watching others (quadrature translates them to apply to oneself), hearing stories (the best stories are chains of causal triples about a character with whom the listener can identify), or even running models in the imagination, "letting our theories die in our stead," in Karl Popper's marvelous phrase.

POWER OF THOUGHT

Wait a minute, you might say. The sigma does this complex operation, to every single memory at once? Does not that let us in for a lot of useless computation? (By the way, when this use-all-the-memories technique, or something like it, is used in AI, it is called case-based reasoning.) For some cases, yes; but for others, that is the cost of generality and flexibility. Beyond that, the brain is not like a von Neumann computer architecture with an active processor on the one hand and a passive memory acting like a filing cabinet on the other—the memories are right there in the neurons that are doing the processing, stored in what form exactly we don't know yet, but the short answer is, "It is all processor." What is more, once we know that the memories are going to be matched in a certain way, it's not a particularly difficult feat of programming to index them so that the bulk of the work of matching is done on the ones closest to a reasonable match.

The other, more interesting part of the answer is that putting memories in is not like stuffing paper into a filing cabinet, either. Memories fade with age, unless reinforced. Memories similar enough merge into generalities (Note that since there are many sigmas specialized for many purposes, memories will be merged and generalized in some but not in others. The ones in your balance control have merged all the steps you have ever taken into a nice smooth control model, but somewhere else, the steps up onto the podium to receive your diploma stand alone.)

This is, of course, where nanotechnology and Moore's Law come in. The processing power of the brain has been estimated at a million operations per second per neuron times ten billion neurons. Using fancy programming tricks we might cut that by a factor of one hundred to one thousand, but probably not much more. That leaves us needing at least a thousand, possibly ten thousand, current-day high- end PCs to run a full fledged AI. Now we have at least one idea about what all that computation might be doing.

It's worth pointing out that some scientific and commercial computer systems have this kind of processing power right now. Of course, they cost tens of millions of dollars. Moore's Law puts this on your desktop for one thousand dollars in twenty-five years.

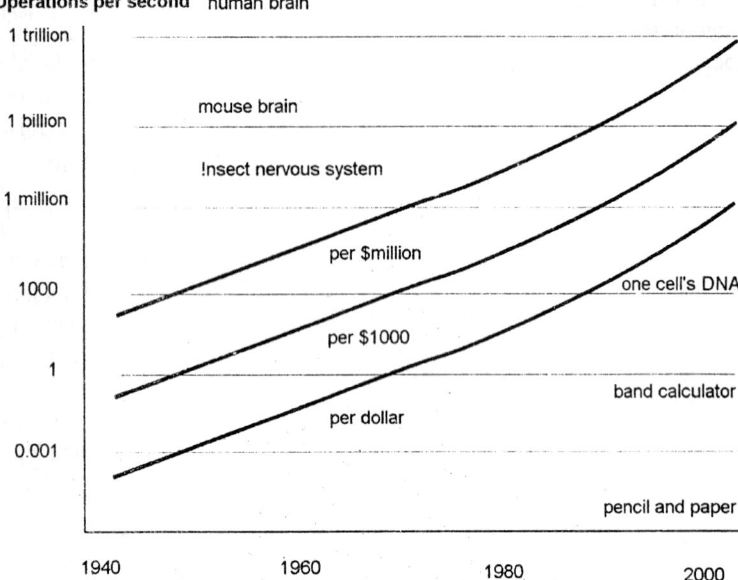

Fig. 6.2. Moore's Law.

So suppose you sell your house and buy one? What could you do with it? If you have an information job, anything from architecture to managing a store, the machine could do it for you (with a few cameras and speakers in the right places). Suppose your job is writing AI soft ware? Your AI proceeds to write a better version of itself, with no further help from you. Better can mean two things, and usually means both: it can do more or be faster with the same hardware or it can take advantage of more hardware are to do even more than that. In either case it's now smarter than you are.

Singularity

The reciprocal of one-tenth is ten. The reciprocal of one-hundredth is a hundred, and the reciprocal of one-thousandth is a thousand. As the number gets closer to zero, the reciprocal rises precipitously. Zero has no reciprocal at all; zero is called a *singularity point* for reciprocals.

Vernor Vinge, computer scientist and a grandmaster of science fiction, coined the term *singularity* to describe a point in the future where the intelligence of machines increases precipitously. It is not likely that it will be a true *mathematical singularity*, which would imply that progress would increase without bound before a specific date. However, as a general term for a time in the future where machine intelligence increases to the point that we humans are unable to keep up, it has caught on as a buzzword among the technophilic avant-garde.

First of all, can this really happen? There are alternate theories: suppose that human intelligence is an optimum point and that adding more processing power hits a law of rapidly diminishing returns. Committees certainly give credence to this idea. It is true that we don't own how to organize a superhuman intelligence, or even a human one; but rapid strides are being made in the area, particularly in understanding how human and animal minds operate. It seems likely that we can do at least a little better once we have a full under standing.

Another indication that AIs could exceed humans is that AIs could have low-level, instinctive access to the kinds of abilities that computers have now, but that humans do poorly, such as solving partial differential equations, maintaining billion-fact databases, and counting the words in this book in approximately a millisecond.

So are we to be outclassed, left in the dust, adopted as pets if we are lucky, and in general suffer a crushing blow to the human spirit because we are no longer the smartest kids on the block?

Of course not. For one thing, we built these machines and we get proud-grandparent credit for anything they can do. (Let's make sure it's credit and not blame we are due) then, we always have the option of improving ourselves as well as the machines, something we'd want to do anyway.

But the most basic reason is that most of us are not the smartest kid on the block to begin with. There is only one per block. Most of us are not haunted by nightmares because we can't beat Deep Blue at chess; we can't beat Garry Kasparov, either. And none of us can beat City Hall.

Synthetic organizations—club, societies, guilds, and institutions—as distinct from autocracies and natural, fluid groups of people, have been around for thousands of years, and nothing is a clearer trend in history than the rapid rise in their size and complexity. It seems a good bet that the first general AIs will show up in the decision making apparatus of corporations, since that's where the need is great and the resources are available. Working for, or trading with, such a machine won't be all that different from dealing with the cur rent paper machine. When it is different, it will in general be better. Corporations want their customers to be satisfied, and they want their employees to be happy—or they'll lose unhappy talent to ones that do.

Another crucial implication of AI is that the design of ever-more-complex machines and systems will continue to be possible. We should expect that we'll be able to have systems more complex than we do now, which work better and are more reliable, and will still be able to afford them.

Suppose that in 2010 a certain system costs $10 million to design. Assume that is when the human and AI designers are equivalent in cost, and the AT designer gets cheaper at Moore's Law rates. Then in 2020 the same system would cost $175,000 to design; by 2030, $3,000, and by 2050, less than $1. Somewhere in between it gets economical to design each individual item, not just for the individual person using it but for the particular instance of use. Not just customized—a whole new kind of machine invented as if a top-notch engineering team spent years working on it, just for your task of the moment.

This would be impossible to cope with if you had to learn a whole new set of techniques, commands, and capabilities each time. But the new device will contain an AI as well as being built by one. It will be an expert at the kind of job it is supposed to do, but not only that; it will know you, understand your style of operation and interaction, and be familiar with the interfaces you are used to. So when you go to the garage to go to Phoenix alone, you find a completely different kind of car waiting than if the whole family were going to Nova Scotia. It might seem overkill to reinvent a whole new kind of vehicle for each trip, but you can see how something like this will come in handy for controlling Utility Fog.

7

DESIGNING COMPUTATIONAL NANOTECHNOLOGY

Using computational nanotechnology to design systems, ahead of our ability to manufacture them in the laboratory, supports developments in the laboratory all the way from the short term through the long term. One of the remarkable characteristics of computer-aided design of molecular machines is that it is actually easier to design advanced systems than it is to design early systems, because the advanced systems can be built with better tools, under fewer constraints. Our speaker on this topic comes from an interesting background. He went from electrical engineering to becoming one of the co-inventors of public key *cryptography*, a crucial development in computer security and communications. He joined Xerox PARC to work on computer security, and somehow got his job description changed to computational molecular nanotechnology.

DIAMOND

If you follow Feynman's idea, you develop a great interest in *diamond*, because diamond is stronger, lighter, has better thermal conductivity, has a wider electron band gap, and so forth. It is the general, all-around wonder material, that there are other arrangements of carbon atoms that also have excel lent properties (such as graphite and more novel graphitic structures like "*buckytubes*"). Fundamentally, if you think about diamond or diamond materials when we talk about things we would like to make and materials that we will be using once we have molecular manufacturing, you will not be far wrong.

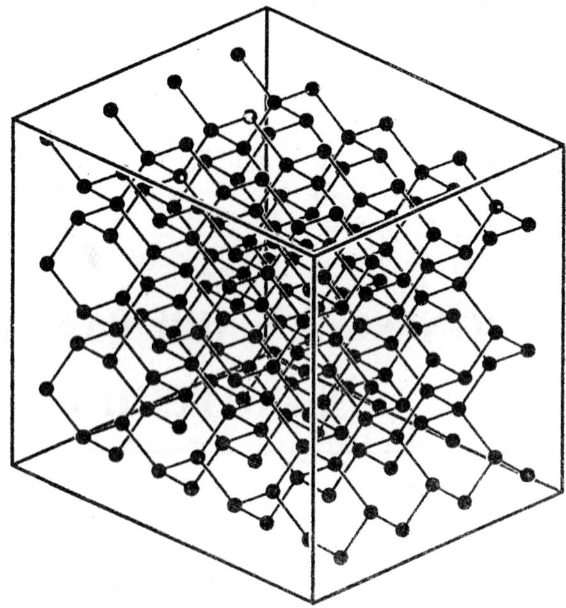

Fig. 7.1. One cubic nanometer of diamond, showing the arrangement of the constituent 176 carbon atoms.

The arrangement of *atoms* in one cubic nanometer of diamond. Each carbon atom has four bonds to its neighbouring carbon atoms, which hold it in place very securely. These carbon-carbon bonds have distinct mechanical properties. The individual bonds have a characteristic length and a characteristic tensile strength. They will stretch a bit if you pull on them, and break if you pull too far. We can think about bonds as little springs.

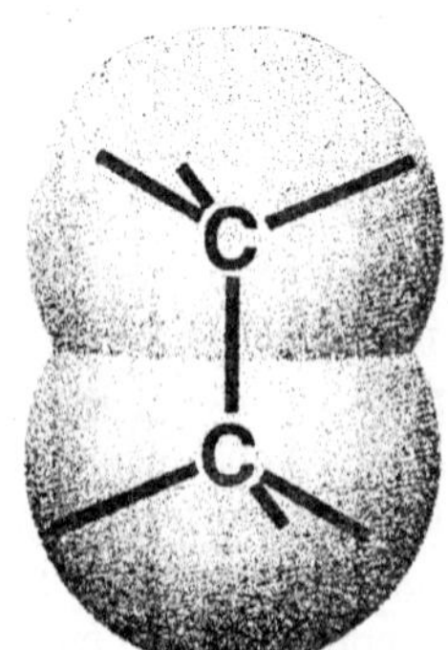

bond length 0.154 nanometers

spring constant 440 newtons/meter

working stress 3 nanonewtons

Fig. 7.2. Mechanical characteristics of carbon-carbon single bonds.

More quantitatively, the energy of the bond as it is stretched can be modeled by the Morse potential. The minimum of the curve shows the optimum length of the bond, about 1.54 Å. As the bond is stretched (which moves us to the right on the curve) the bond resists, since it wants to return to its equilibrium length. The Morse potential shows that the carbon-carbon single bond actually can be stretched quite a bit. The inflection point in the curve is not reached until the bond has been stretched about 22 percent of its original length. Because you can stretch this bond substantially before it breaks, it is possible to do things that are moderately surprising, compared to our expectations of diamond. It is possible to build structures that contain carbon-carbon bonds that are fairly highly strained.

Designing Small Mechanical Structures

Because of the tolerance of these bonds to strain, you can imagine structures in which a thin, flat sheet of diamond is bent into a hoop. Bending the diamond sheet to form the outer sleeve of this bearing imposes a considerable strain on the car bon-carbon bonds, although not enough to snap them. The surfaces are finished by attaching other atoms so that no carbon atom has "*dangling*" bonds (that is, bonds

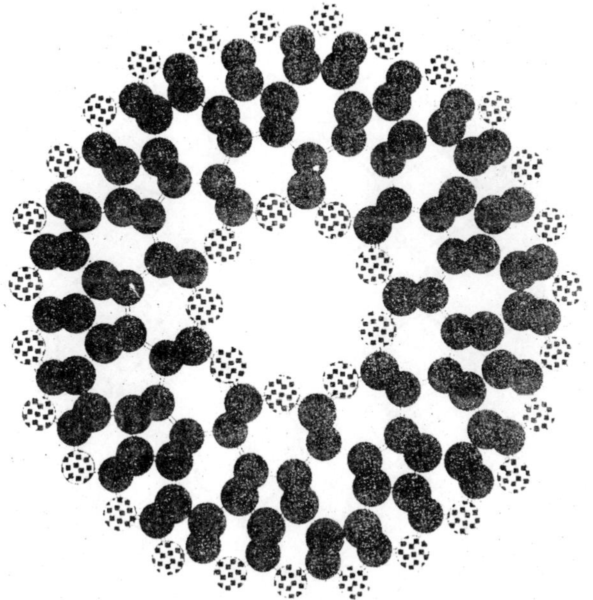

Fig. 7.3. A thicker sheet of diamond, bent by the introduction of dislocations, yields a shaft.

that are not used to attach to other atoms). The inner shaft of the bearing is similarly formed of a hoop of diamond with the surfaces finished with atoms that provide a stable surface structure. Placing the shaft within the sleeve forms a small, but quite functional, bearing.

To bend a sheet of material into a hoop, it is necessary that the sheet be thin. Attempting to bend a thick sheet of diamond into a hoop will create too much strain and the diamond will break. A somewhat thicker sheet of diamond is bent into a hoop, but to prevent the diamond from breaking, the strain has been relieved by introducing dislocations into the diamond structure. These dislocations include irregularities in the crystal structure, such as five-member and seven-member rings of carbon atoms. The specific dislocation that we have introduced is called the *Lomer dislocation*, and was described by Lomer in the 1950s. A single Lomer dislocation which the regular crystal pattern of the diamond (six-member rings of carbon atoms when viewed in a flat projection) is interrupted by a seven-member ring adjacent to a five-member ring. The Lomer dislocation essentially puts a small bend into the diamond sheet.

Introducing many Lomer dislocations allows bending the diamond sheet into a tube. Bending another sheet around the first tube gives a

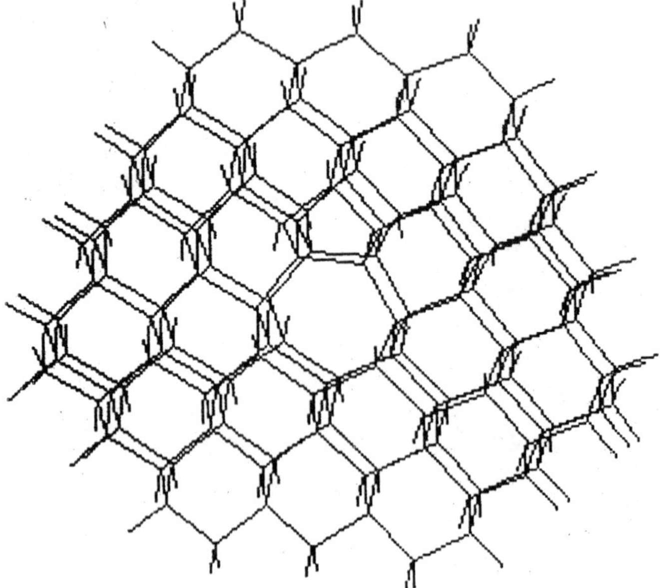

Fig. 7.4. The arrangement of carbon atoms in a single Lomer dislocation.

very nice molecular bearing that involves a much thicker wall than the bearing. In principle this strategy could be extended to produce as thick a tubular structure as desired.

Properties of Molecular Structures

A commercially available computational chemistry package was used to do the energy minimization of the bearing—PolyGraf from Molecular Simulations, Inc. (MSI), which uses the Dreiding II force field, a general-purpose and flexible force field for molecular mechanics.

The simulation shows the behaviour of the bearing at zero Kelvin while being rotated very slowly. We do this by minimizing the energy of the structure, rotating the shaft a little bit, minimizing the energy of the new structure, rotating the shaft a little more, and so on. Observing an animation of the series of structures produced by these rotations and minimizations gives the appearance of the shaft turning smoothly in the sleeve. You will notice that the outer sleeve appears completely immobile. It is actually moving very slightly, but the movement is too small to be seen in this video.

A smaller bearing containing fewer atoms (but showing the same properties) was designed by Eric Drexler. Again, an animation showing the bearing rotating very slowly at zero Kelvin shows the central shaft rotating smoothly with very little other atomic motion. "Slowly" in this context means slow in comparison with the native vibrational frequencies that occur in this structure. This structure is very small, so the vibrational frequencies would be in the terahertz range. Accordingly, the slow rotation might occur at gigahertz frequencies. "*Slow*" is a relative term! The hydrogen atoms on the outer sleeve can be observed to wiggle a bit back and forth as they adapt to the rotation of the shaft. This is not a vibrational motion that occurs (because of the slow speed of the rotation). It is simply that as the shaft rotates, the hydrogen atoms wiggle back and forth to minimize the energy of the structure.

The energy barrier for rotation of the shaft inside the sleeve is very small, over 1,000 times smaller than kT (thermal noise) at room temperature. The larger bearing was designed at PARC using software written by a summer intern from MIT, Lakshmikantan Balasubramaniam. It was designed using dislocations in the diamond structure of the kind. The shaft fits snugly into the sleeve and rotates smoothly.

The two surfaces have periodic bumps and hollows, but the periods are different for the two surfaces—the two surfaces are incommensurate.

This means the two surfaces can not "*lock up*" in any particular position and makes the barrier to rotation very low. In this case, we made the two surfaces incommensurate by using symmetry. The symmetry of the inner shaft is nine fold. That of the outer sleeve is tenfold. The numbers 9 and 10 are relatively prime (they have no common factors other than 1), so the bumps and hollows on the two surfaces cannot lineup. This same concept has been used in the other bearings and can be used more generally to make two surfaces slide over each other very smoothly.

Animation

Animation of the computer simulation shows that the central shaft seen in the figure rotates rapidly, while the output shaft rotates slowly. The small planetary gears rotate around the central shaft, and they

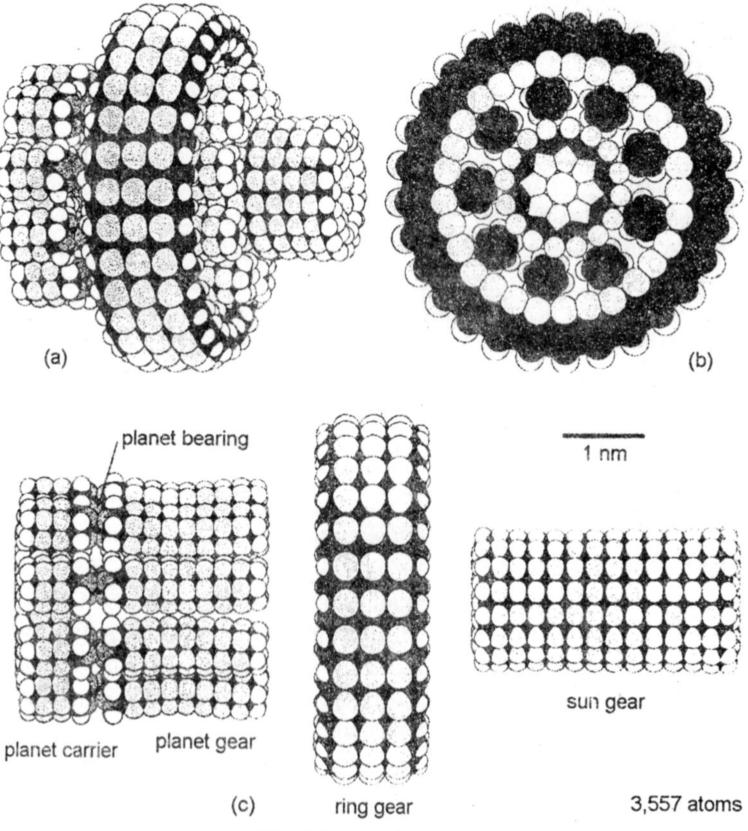

Fig. 7.5 Planetary gear.

are surrounded by a ring gear that holds the planets in place and ensures that all of the components move in the proper fashion. The animation shows that the ring gear wiggles alarmingly because it is rather thin. In an actual system, the ring gear would be part of a larger wall that would hold it solidly in place and eliminate the alarming motion seen in the animation. Nevertheless, the simulation illustrates a simple mechanical device. Planetary gears are used in automobiles and other places where you want to transform speeds of rotating shafts. The only thing that distinguishes this planetary gear from other planetary gears is that it is remarkably small.

Designing a Nanoscale Robot Arm

The conclusion drawn from this work is that the conventional mechanical devices (that we are familiar with at a macroscopic size) can be scaled down to the molecular size range and still work, at least if we are careful about the design. We should be able to build a wide range of devices that use molecular tubes, bearings, gears, shafts, and so forth. This device has about 5 million atoms, so it is a bit too big for us to model on a computer today. It should not be too long, how ever, before we have the tools to model it, as we have done with the planetary gear and other structures.

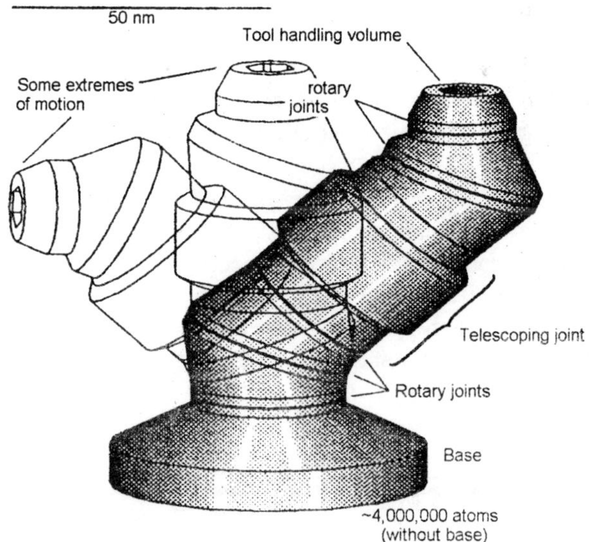

Fig. 7.6. Schematic drawing of a rigid nanopositioning mechanism.

This arm is designed by using the same ideas as in the simpler structures—solid diamond for high strength and stiffness, the introduction of dislocations to relieve strain when we bend the material, incommensurate surfaces at the joints so they will move smoothly and easily, and so forth.

Chemical Tools

So far we have concluded that if we could arrange the atoms the way we wanted, we could build some very nice devices, including small robotic arms. But how do you rearrange atoms? How do you synthesize things like those devices? We could start by looking at how diamonds are currently synthesized. Today, diamonds are often synthesized by the use of *chemical vapour deposition* (CVD) methods.

CVD methods are fairly straightforward conceptually. You take a flat sheet of diamond and pass a reactive gas over it. The reactive gas contains atomic hydrogen, which bumps into the diamond surface and activates it. The diamond surface is normally passivated (that is, it has hydrogen atoms on the surface). Carbon atoms on the surface do not have dangling bonds, because the hydrogen on the surface terminates their bonds. Atomic hydrogen can react with the hydrogen that is bound to the surface, and strip it off. This leaves a dangling bond on the surface, which is very reactive. This process is called "hydrogen abstraction" and is a commonly postulated first step in the growth of a diamond film. To grow something on the surface, you must first make it active. To make the surface active, you remove a hydrogen atom and leave a dangling bond on a carbon atom. The place where you removed the hydrogen atom is now very reactive, so a molecule containing carbon can now react at that spot on the surface and make the surface grow.

Of course, if we want to make something where every atom is in the right place, we can not rely on a reactive gas that randomly reacts with any spot on the surface. We need tools that we can position and that will react at exactly the spot we want them to. It would be used as one of the first steps in site-specific synthesis of a diamondoid structure. Illustrated is a surface that has a hydrogen atom.

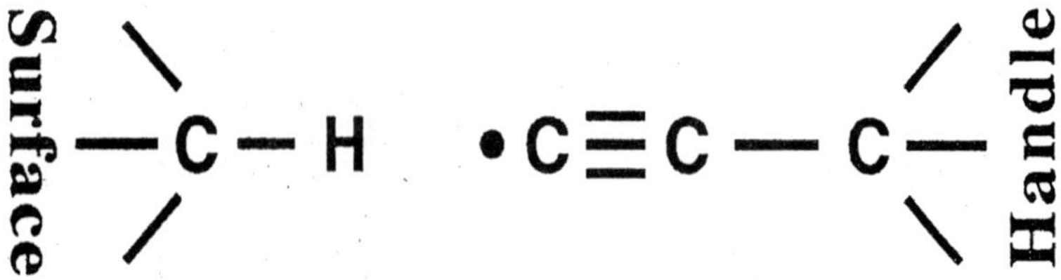

Surface —C—H •C≡C—C— Handle

Fig. 7.7. A site-specific hydrogen-abstraction tool.

The hydrogen-abstraction tool has a handle that might be held by the robotic arm. The tool approaches the surface so that it is positioned directly over the specific hydrogen atom that we want to abstract. The triple-bonded carbon at the tip of the abstraction tool has a dangling bond and a very high affinity for hydrogen. The bond between a hydrogen and a carbon atom that is triply bonded to another carbon atom is stronger than almost any other bond that hydrogen forms. The hydrogen atom on the surface has the "choice" of being bound by a reason ably strong bond to the surface, or to the very high affinity site on the tip of the abstraction tool. So, it "decides" to be on the tip of the abstraction tool instead of on the surface.

Hydrogen Abstraction

The *hydrogen abstraction* from the diamond surface has been modeled by Charlie Musgrave and Jason Perry at Cal Tech using higher-order *ab initio* calculations to analyze the barrier height to this reaction. They found that the barrier was either very low or zero. The low barrier means that if you bring this tool up to any point on a diamond surface that has a hydrogen atom sticking out, you can pluck that particular hydrogen atom off the surface.

This is an illustration of the kind of atomically precise reactions that would be used in building diamondoid structures. Obviously you would also require tools that deposit carbon atoms, so that after removing a hydrogen atom, you could move in with a second tool and use a carbon-deposition reaction to continue building the structure. There are several proposals for carbon-deposition tools.

The *hydrogen abstraction* tool poses an interesting problem: If you chose the tool so that it has a greater affinity for hydrogen than almost any other structure, how do you get the hydrogen off the tip? There

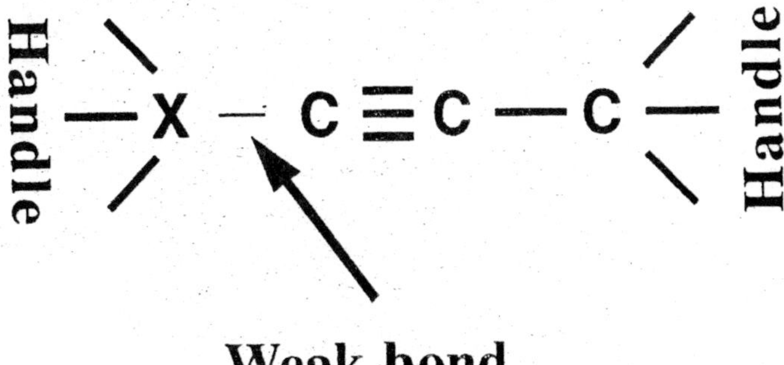

Fig. 7.8. Precursor to a hydrogen-abstraction tool.

are several answers. The simplest answer is that you do not—you throw out the tool. To make that answer work, you need a precursor to the tool. One possible precursor that has two handles and a bond designed to be deliberately weak. "X" might be Ge or Si. This structure is reasonably stable, but when the handles are pulled apart hard enough, something has to give. The deliberately weak bond will break, giving you the hydrogen-abstraction tool. This tool is used to abstract a hydrogen atom and is discarded after use. Another precursor tool is used for the next step, activated by snapping the weak bond, the next hydrogen atom is plucked from the work surface, and this tool is then also discarded. This cycle is repeated several times. In this particular approach, the device is fed a steady diet of hydrogen-abstraction tool precursors.

Synthetic Strategy

These considerations taken together suggest the following synthetic strategy for building complex, atomically precise, diamondoid materials. The first feature of the strategy is positional control, a robotic arm that can position molecular tools. The second is the use of molecular tools derived from highly reactive compounds, such as radicals, carbenes, and so forth. And thirdly, the synthesis is done in an inert environment (for example, a vacuum) so that there will be no side reactions between the highly reactive compounds and the environment of the synthesis.

The basic features of the synthesis are thus a molecular work piece that is being built, a molecular tool that has a highly reactive chemical species at its tip, and an arm to position the molecular tool. The arm pushes the tool against the work piece and then pulls it away. Because the tip of the tool is a highly reactive species, some reaction will occur at the surface of the work piece. A series of site-specific reactions can then be used to build the desired structure.

To make this strategy work, we will need a complete set of molecular tools, each one of which can be used to drive a particular chemical reaction. Perhaps 20 or so different tools might be needed, each one of which would be used to make a specific modification of the work piece. The set of chosen tools is then used to build complex, atomically precise structures. Such a strategy should permit the construction of a broad range of diamondoid structures.

Self-replicating Systems

This general field of self-replicating systems has been studied reasonably extensively. A NASA study found rather interesting was

"Advanced Automation for Space Missions." Several hundred pages were devoted to the Self-Replicating Systems (SRS) concept. One of the conclusions that they reached was the following:

"The theoretical concept of machine duplication is well developed. There are several alternative strategies by which machine self-replication can be carried out in a particular engineering setting."

The archetypical proposal for a self-replicating system is the von Neumann architecture that couples a universal computer (which can compute anything you want) with what von Neumann called a "*universal constructor*" (which can construct anything that you want). Von Neumann's detailed analysis of a universal constructor was done in a theoretical, two-dimensional, cellular automata world. In this theoretical two-dimensional world, he had an arm that would move about and would change the state of the cell at the tip of the arm. This allowed you to build devices in this two-dimensional world.

To build things in our three-dimensional world, von Neumann's architecture must be modified along the lines of Drexler's architecture for an assembler. This architecture employs a (very small) molecular computer and a "*molecular constructor.*" The constructor consists of two major components. One component is a molecular positional capability, like the robotic arm (although many other positional mechanisms are possible). The other component is a well-defined set of reactions that occur at the tip of that arm (such as the hydrogen-abstraction reaction). These components together comprise a general-purpose device that can make a wide variety of interesting structures, including copies of itself.

Starting with Simpler Systems

Building the first assembler will probably be hard, so we would like to simplify the problem as much as we can before we start. As a consequence, one design objective is to think about devices that are not general-purpose assemblers in the sense of the von Neumann architecture, but nonetheless provide the ability to make a wide variety of structures. One of the first things that can be done is to adopt the broadcast architecture. This simpler architecture uses a macroscopic computer to do all of the computation. The "*stripped down*" assemblers consist of only a molecular constructor. The computational element has been moved into a large, macroscopic computer where the computation is done once, and the simple instructions that result from the computation are broadcast to the actual constructors. The molecular constructors are thus "*gutless wonders*," they are barely able to receive and interpret some simple instruction, such as "move your arm one nanometer to the left." This architecture allows the molecular constructor to be simpler and smaller than a general-purpose assembler,' while the computational device is a conventional macroscopic computer.

Several specific advantages result from' this architecture. First, the smaller molecular constructors contain fewer atoms than a general-purpose assembler. Second, these devices are much reduced in complexity (and thus easier to design) because they contain fewer components. Third, molecular constructors are easily reprogrammed, which is very important for use in manufacturing because the same devices can be reprogrammed to build a broad range of different products. Finally, this design is inherently safe. If a macroscopic computer is issuing a stream of instructions to molecular constructors inside a vat, then each of those constructors will be unable to do anything when cut off from the broad cast instructions. They lack the capability for independent self-replication. They require an environment where they have broadcast instructions.

In summary, the broadcast architecture is economically more attractive, simpler to design, and more safe. It is, therefore, an interesting target for early designs.

Different Kinds of Self-replicating Systems

Many people object that self-replicating systems must be extremely complicated. That is not necessarily so. Von Neumann's universal constructor is about 500 kilobits and has been partially simulated on computers. The Internet Worm, a self-replicating piece of software that replicates in an artificial environment, is also about 500 kilobits. The bacterium *Escherichia coli* is certainly an authentic self-replicating device that makes copies of itself in the real world, and it has a complexity of about 8 megabits (the size of its genome), which equals 1 MB and fits comfortably in the random-access memory (RAM) of most personal computers today, so it is not really all that complicated. Drexler's assembler has an estimated complexity of about 100 megabits. That translates into about 12 MB and would fit into the RAM of many higher end personal computers.

Human beings have a complexity of about 6.4 gigabits. That is rather substantial. However, human beings do things besides just replicating, so it is unfair to attribute this complexity just to self-replication. The NASA proposal for a lunar manufacturing system was for a very complex system. They proposed to put a cross section of

Table 7.1. Complexity of self-replicating systems

Self-Replicating Systems	*Complexity in Bits*
Von Neumann's Universal Constructor	About 500,000
Internet Worm	500,000
Escherichia coli	8,000,000
Drexler's Assembler	100,000,000
Human	6,400,000,000
NASA Lunar Manufacturing Facility	More than 100,000,000,000

earth-based manufacturing technology on the moon, set up a mining facility to mine lunar soil, and smelt and refine that soil to build other components. They estimated the complexity of that system at more than 100 gigabits. They were really pushing the limits of what today's technology can do in terms of design complexity. Nonetheless, they concluded that with sufficiently generous funding such a system could be built within about 20 years.

We find that Drexler's assembler is somewhere in the middle of this range of complexities. Furthermore, the original design of Drexler's assembler can be simplified. Right now, it is not entirely clear how simple we can make an assembler and still have it work. This is an area worth very serious research. If we can make such a device simple enough, it will be easier to make.

Reactions to Molecular Manufacturing

How are people reacting to these ideas? The following is an interesting quotation from the corporate world: "I believe that nano science and nanotechnology will be central to the next epoch of the Information Age, and will be as revolutionary as science and technology at the micron scale have been since the early '70s indeed, we will have the ability to make electronic and mechanical devices atom-by-atom when that is appropriate to the job at hand."

Advances in Computer Technology

You need some sort of molecular computers to run your molecular manufacturing capabilities, so you are going to have to shrink your computers to a small size. The trend toward smaller computers has been fairly steady for the last 50 years. In 1950, it took 10^{19} or more atoms to store one bit. During the 1960s and 1970s, it decreased to 10^{13}, and by 1990, it was down to 10^{9}. If you extrapolate this into the twenty-first century, you find that sometime between the years 2010 and 2020, you are using one atom to store one bit of information.

If you look at the finest machining technology available, you find that the trend since the turn of the century has gone from about 10 microns to a few nanometers today, again a straight line on a semilog plot. This trend reaches atomic diameters sometime around the year 2017.

If you look at the energy dissipated per logic operation, you find a similar trend. There has been a steady decrease, starting with relays in the 1940s, then moving on to vacuum tubes at 10^9 picojoules per logic operation, then on to transistors, integrated circuits, and today large-scale integration and ultra-large- scale integration. Around the year 2015, this extrapolation reaches kT at room temperature. This is roughly the energy of a single atom bouncing around at room temperature. This is not very much energy. If these trends continue, then sometime around the year 2015, we will be building computers that use one atom to store one bit, are built with atomic accuracy, and dissipate about kT per logic operation.

These extrapolations are one way to answer the question of how long it will take to develop molecular manufacturing. While there is no way to guarantee these trends will continue, they have held steady for the last 50 years through dramatic changes in technology. It would be prudent to be prepared if they continue on schedule.

Fundamental Physical Limits of Computation

Such arguments have people thinking about the fundamental physical limits of computation. How far can we carry the process of decreasing the size to store a bit or the energy to perform a logic operation? The answer appears to be "Quite far!" Computers are built out of atoms, so that is one limit. Computers cannot operate with signals faster than the speed of light, so that is another limit. There do not seem to be any other fundamental limits clearly visible.

People have started to get interested in molecular electronic devices because molecular electronic devices are marvelous and seem to allow us to do man wonderful things. People have also started talking about atomic switches including the work of Don Eigler at IBM Almaden Research Center. These workers have built an atomic switch that tosses a single xenon atom back an forth between a surface and an STM tip. That is indeed an atomic switch although if you count the many atoms involved in the support structure, the switch is rather larger than a single atom. Nevertheless, people are now thinking in the direction of very small switches and saying that it should be possible to build something like that.

MOLECULAR MECHANICAL COMPUTER

The idea of mechanical computers was first proposed by Charles Babbage in the 1800s. His "Analytical Engine" included a central processing unit (CPU), memory, and stored program control. It illustrates quite nicely that mechanical computers are quite feasible.

One approach to molecular mechanical computation is to have rods sliding back and forth with knobs sticking out from the rods. The basic concept is very simple. Two objects cannot occupy the same place at the same time. If you have one rod sliding back and forth in one direction with a protruding knob, and a second rod sliding back and forth at right angles to the first, also with a protruding knob, then the two knobs can bump into each other. If the first rod slides to one side, then the second rod can now move freely. If instead the first rod blocks the motion of the second rod, then the second rod cannot move because the knob on the second rod is blocked by the knob on the first rod. That simple blocking interaction can be used as the basis for logic in mechanical computers.

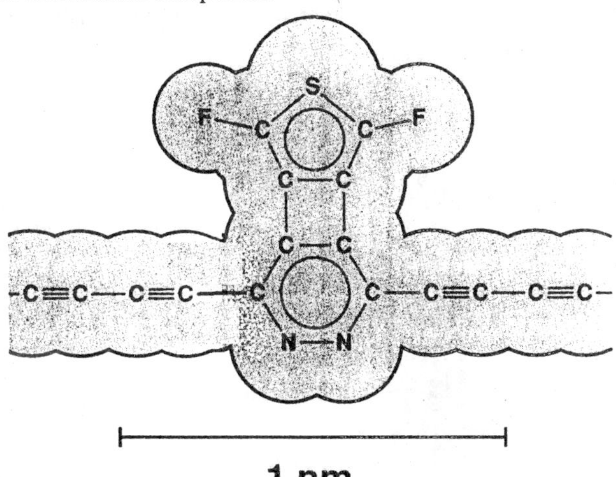

Fig. 7.9. A proposed carbyne rod component for a molecular scale mechanical computer.

One rod that can slide back and forth and either blocking (0) or not blocking (1) the sliding motion of another rod. This is simple NAND gate that uses this blocking and non blocking concept to allow two inputs to be either pulled on (thus moving the blocking knob aside) or not moving it aside if either rod is not pulled on. If the blocking knobs are both pulled aside, then the vertical driver rod can move freely, and the proper logical outputs of "and' s" and "or' s" can be

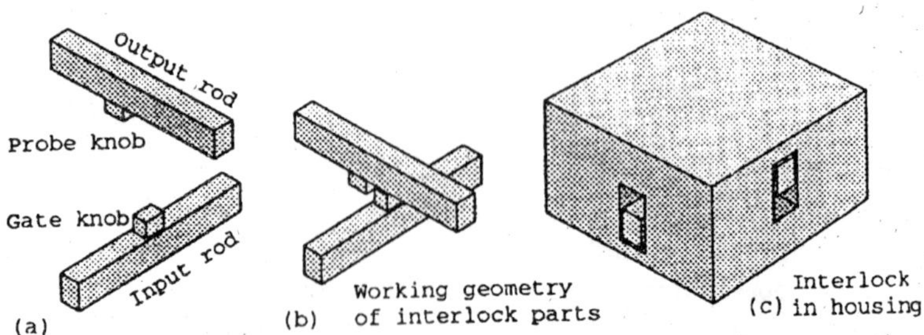

Fig. 7.10. Cross section through a logic gate in Drexler's proposal for a molecular mechanical computer.

computed. As any good computer scientist knows, given a NAND gate, you can build a computer.

The size scale of a computer made out of molecular mechanical devices. The equivalent of a 4-bit microprocessor used in early generation pocket calculators is only a few tens of nanometers on a side, somewhat larger than a ribosome. It would fit very comfortably inside the nucleus of a human cell. Quite remarkably small computers can be made using this approach.

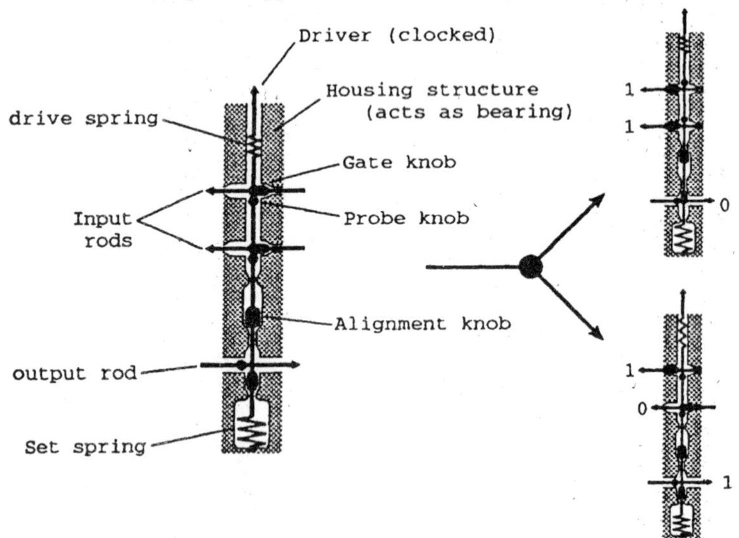

Fig. 7.11. NAND gate in a molecular mechanical computer.

Performance Characteristics

The performance of the computer Drexler has designed can be analyzed. The entertaining thing about the mechanical approach to molecular computers is how good it is. You can get switching speeds of 50 picoseconds with very low energy dissipation. If you are willing to increase the energy dissipation, you can get higher speeds of operation. Conversely, slowing the speed of operation gives less dissipation. At 100- picosecond switching speeds, the energy dissipation would only be of the order of 10^{-23} joules per logic operation. That is a very small amount of energy, less than the energy involved in a single atom bouncing around at room temperature. Until the last year or two, it was not clear how to make molecular electronic devices that dissipated that little energy.

Table 7.2. Rod logic parameters for molecular mechanical computation

Length	50 nanometers
Displacement	0.84 nanometers
Travel time	50 picoseconds
Peak speed	35 meters per second
Mass	1.6×10^{-23}
Force	3.4×10^{-11} Newtons
Acceleration	2.1×10^{12} m/sec^2
Peak kinetic energy	9.8×10^{-21} joules
Dissipation	$<4.0 \times 10^{-21}$ joules

The storage capacity of a computer system built using this mechanical logic proposal is very high (10^{20} bits per cubic centimeter).

Table 7.3. Storage capacity of a molecular mechanical memory system

Volume per bit	10 cubic nanometers
Access time	1 nanosecond
Storage capacity	10^{20} bits/cubic centimeter (cc)

Such a computer could contain 10^{12} CPUs per cubic centimeter. It turns out that the major limiting factor in the operation of such a computer is the power dissipation. By limiting ourselves to a mere 10^{24} logic operations per second, we can keep the heat dissipation down to a manageable 7 kilowatts or less. Water cooling can handle this. The amount of water would be comparable to the output of a conventional faucet. For a more conservative design, the 10^{24} logic operations per second could be decreased to merely 10^{21} operations per second, thus decreasing energy dissipation to only 7 watts. If you are willing to tolerate a comparatively underpowered computer (only several orders of magnitude more powerful than a Cray), then the energy dissipation can be made smaller.

Table 7.4. System parameters for molecular mechanical computation

Volume	1 cubic centimeter
CPUs	10^{12}
Gate operations	10^{24} per second
Power dissipation	7000 watts
Cooling water	70 cc/second
Temperature change of coolant	25° Kelvin

Advantages and Disadvantages of Molecular Mechanical Logic

The first advantage of molecular mechanical logic is that it is conceptually simple. It can be explained to a wide variety of audiences.

Second, there are many possible realizations of this proposal. Since the design is fundamentally mechanical, details of how the rods and knobs are designed can be changed (if some aspect of the design turns out to be objection able) without invalidating the entire design. For example, the rods could be made slightly larger and thus slightly slower, if necessary, but the revised design would still use the same concept. Thus, the design is immune to minor bugs. If a particular structure does not work, there are a broad range of other, similar structures that should.

Thirdly, the mechanical proposal scales from the macroscopic to the molecular in size. This is an advantage that electronic devices might lack, primarily because of problems introduced by electron tunneling at small scales. Quantum mechanical tunneling is not a problem with molecular mechanical designs.

Finally, molecular mechanical devices can be made thermodynamically reversible. This means that very low energy dissipations are possible. We have only recently discovered how to make thermodynamically reversible systems using conventional electronic devices. The major disadvantage of mechanical logic is that it is almost certainly going to be slower than future electronic systems. This is because nuclei are more massive than electrons, so a device that depends on moving nuclei will necessarily be slower than a device that depends on moving electrons. Therefore, even though switching speeds of 50 picoseconds should eventually be possible with mechanical logic, it is hoped that in the future, electronic devices will be much faster. Rod logic might be seen in the future as historically interesting, but will be beaten by other approaches. On the other hand, it looks like it will be hard to beat the molecular mechanical logic designs for small size. High- density memory, or systems where small size (and small atom count) are at a premium, might well be made by using molecular mechanical methods even in the future.

Research Objectives for Molecular Manufacturing

The major research objectives in molecular manufacturing are to (1) design an assembler, (2) computationally model it, and (3) build it.

One approach to building an assembler is shown in Fig. 3.12. The synthetic capabilities that we have today are portrayed by the point on the left of Fig. 3.12. As we go forward in time and our experimental capabilities increase, our range of synthetic capabilities also will

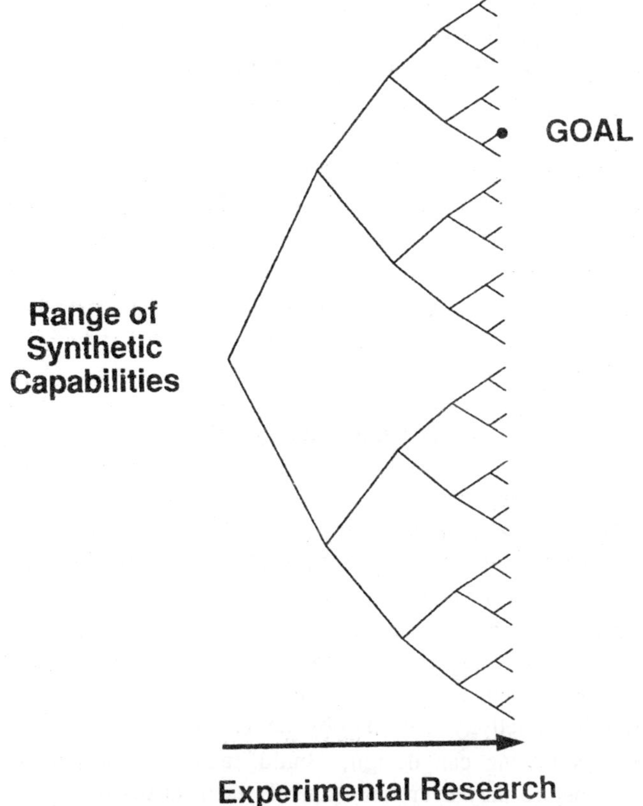

Fig. 7.12. A schematic illustrating how experimental progress will increase our range of synthetic capabilities, leading to the goal of building an assembler.

expand, so that we will be able to build a wider range of products. If the goal is to build an assembler, the expansion of our general ability to build products eventually will expand to the point of including the ability to build an assembler.

Here, the physical experiments would be combined with computational experiments, using computational experiments to move backward from the objective, to eventually link up with the forward progress in experimental technology. In this way, you are simultaneously trying to expand experimental capabilities and to increase your understanding of what the goal looks like through computational modeling. This process should help to identify the most attractive experimental paths to reach the goal.

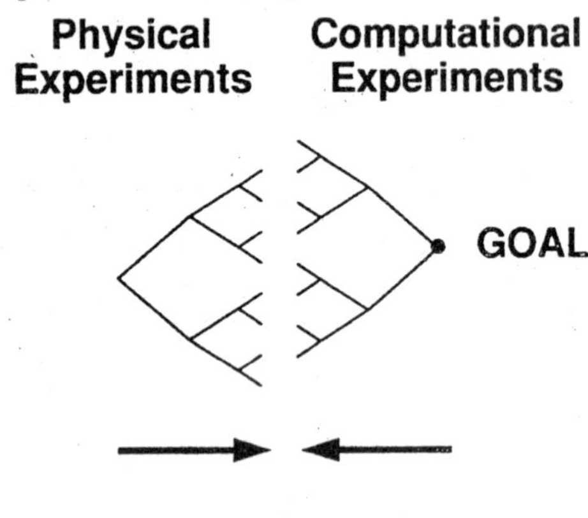

Fig. 7.13. A more efficient road to building an assembler by cooperation—combining physical experiments and computational experiments.

This combined approach will shave a great deal of time from the development cycle for building an assembler. One observation is that there is a lot of work being done with physical experiments now, and very little work being done to computationally model molecular manufacturing systems. There is a great deal to be gained by pursuing work in computational modeling, simply because few people are doing it, and it is an activity that has obvious long-term value.

Just as Boeing can design, "build" and "fly" airplanes on a computer before they ever build and fly them in the real world, we can model the molecular manufacturing systems of tomorrow on our computers today. While it is hard to say how long it will take to develop molecular manufacturing, it is clear that we can develop better systems more quickly if we combine experiment with theoretical and computational approaches. Each offers unique advantages, advantages that are complementary and can be combined into a better development path than either alone.

The design and fabrication of the first molecular manufacturing systems will require a broad range of expertise in many different areas. These will include physics, robotics, chemistry, surface science, materials science, computer science, electrical engineering, mechanical engineering, computational chemistry, and others, as well. We are talking about the design and construction of a complex device that will require substantial multidisciplinary work.

8

NANOMEDICINE

It was 1847, less than two lifetimes ago. In the General Hospital of Vienna, one of the leading centers of science and medicine of the day, there were two maternity wards. Mothers in First Ward were attended by doctors and medical students. Mothers in Second Ward had to make do with midwives. Yet the women of Vienna begged to be admitted to Second we since First Ward had a reputation as a deathtrap.

The reputation was true. More than 12% of mothers in First Ward died of puerperal fever after childbirth, while less than 3% of those in Second Ward did.

A brilliant young Hungarian doctor, Ignaz Semmelweis, had recently been put in charge of the maternity wards and was desperate to find the cause of the deaths. He wrote:

Almost every day, the sound of bells intimated that the priest was administering the last sacrament to the dying. I myself was terror-stricken when I heard the sound of bells at my door This worked on me as a fresh incentive that I should, to the best of my ability, endeavor to discover the mysterious agent, and a conviction grew day by day that the prevailing fatality in First Ward could in no wise be accounted for by the hitherto adopted etiology of puerperal fever.

Semmelweis tried everything he could think of. The prevailing medical theories involved things like "miasma" (bad air) and imbalance of "humors" (bodily fluids). Miasma did not work: the air in Second Ward was stuffier than in First. It was more crowded, given the women's quite justified preference. At one point he even hushed up the priests on the theory that the sound of the bells might be contributing. Finally he got a clue when a colleague, Jacob Kolletschka, died from a fever after cutting himself during an autopsy. Semmelweis noted that Kolletschka's symptoms resembled those of the dying women. He guessed that some infectious agent was being carried on the doctors' hands from the autopsies and educational dissections into the maternity ward.

Semmelweis instituted a policy that doctors and medical students wash their hands in chlorinated water before examinations and deliveries. The death rate plummeted to below 2%. The head of the hospital, Dr. Johann Klein, was stung by Semmelweis's insistence that the doctors had been causing the deaths and refused to believe his new theory. The feud became bitter, and the politically connected Klein ultimately forced Semmelweis from Vienna. Semmelweis returned to his native Budapest, where he joined St. Rochus Hospital and repeated his dramatic success against puerperal fever.

In the 1860s a similar controversy raged in the pages of the British medical journal the *Lancet*, as one Dr. Joseph Lister attempted to persuade the medical profession to use carbolic acid as a disinfectant in the treatment of open wounds.

As late as 1879, it was still possible for a distinguished doctor to address the Paris Academy of Medicine and attempt to heap scorn on the idea that infection could be carried on doctors' hands from one patient to another. On this particular occasion, however, the doctor was interrupted from the audience by another who sprang to the podium and drew pictures of streptococci, as seen under the microscope, on the blackboard. The interrupter's name was Louis Pasteur.

By the turn of the twentieth century the germ theory of disease and infection had been largely accepted, and there was a revolution in surgery. Along with antisepsis had come anesthesia. A patient could now enter surgery with a reasonable expectation of surviving it.

By 1950 another revolution had occurred. Sulfonamides, penicillin, and streptomycin had been discovered and put to widespread use. In 1890 the leading cause of death had been tuberculosis, with 245 people out of each 100,000 dying of it each year. By 1990 the corresponding figure was less than one.

A Molecular War

In 1928, when Alexander Fleming discovered penicillin, neither he nor anyone else understood how it worked. Not until later, when the molecular machines that make cells work were being studied, was a full understanding reached.

What happens is this. Bacteria have a tough cell wall, essentially a skin, that is constructed of large, complex, interconnected molecules. To build the wall, the bacterium creates long molecules of saccharide subunits, hangs an amino acid subunit off each one making a comb-like shape, and knits the combs together into a netlike structure called peptidoglycan. The molecular machine that performs this last step is called transpeptidase. The penicillin molecule blocks transpeptidase by getting wedged into its active site, preventing it from working. You can think of this as similar to what happens to a stapler when you try to staple a piece of taffy. When there is enough penicillin around, all the bacterium's transpeptidase "staplers" get jammed, and it cannot join up its peptidoglycan nets to make its skin. So when it tries to divide, a process requiring lots of new skin, it ruptures.

We can flood our bodies with penicillin because the cells of higher animals do not use peptidoglycan, and thus don't have any transpeptidase to block. Bacteria, on the other hand, can become resistant to penicillin by evolving an enzyme that breaks it down. Then we can attack them by adding another drug that inhibits the new enzyme, and so back and forth.

This is how most modern drugs work: by being molecules of just the right shape to form a monkey wrench in some specific molecular machine. We are lucky that there are enough differences in the workings of humans and bacteria to allow us to bollix them up without doing the same to ourselves. Today's medical research studies the molecular machines and reaction pathways that form the workings of cells, and invents new molecule shapes specifically for a particular effect instead of just trying random chemicals to see what works.

Drugs have a couple of major drawbacks, however. They are good at blocking molecular machines; it is harder to fix one that is not working right. This is fine as long as the disease is caused by some bacterium or other form of life that the drug acts as a selective poison for. But having cured many of those, we are left with the diseases that occur when our own machinery goes awry. In 1890 heart disease and cancer killed 170 people per 100,000 per year.

Mind you, the people who die of heart disease and cancer today are dying much later in life, on the average, than the similar number who died of tuberculosis and pneumonia a century ago. But the basic difference is that it is a lot easier to kill it than to fix it, especially when the thing to be fixed, the human cell, is literally a thousand times as complicated as the one, the bacterium, to be killed.

There are, of course, some "fix it" or "maintain it" things that can be done in a drug like way. Vitamins, minerals, and similar nutrients can provide small molecules that the cell cannot manufacture or is not manufacturing enough of. They can provide odd elements, like the iron in hemoglobin, in a form the metabolic machinery can access. Hormones, small drug like molecules that turn on and off various molecular-level functions, can be provided. Insulin is an example. (Drugs that block hormone activity by jamming their receptors are also common.) But you can't take fully formed biological molecular machines as medicine: your digestive system breaks them down just like any other protein you eat.

SURGERY

The wing of medicine that does "fix it" nowadays is surgery. Surgery is quite amazing in its capabilities: reconnecting severed limbs and transplanting organs are not uncommon. For some applications, notably the joints, arthroscopic surgery can be done. The arthroscope is a pencil-sized device that acts as lights and camera for the surgeon to operate through small incisions. This can enormously speed healing times as compared with techniques with large incisions and direct observation.

Nanotechnology will advance the arthroscope in obvious ways. It should be possible to do some kinds of surgery by injecting a thread no wider than a hair. The thread would be a conduit for a vast army of nanomachinery that would build it out into a network around the region of interest and proceed to reconstruct the diseased or damaged tissue. By and large, with this technique, the nanomachines would not add and remove material as much as is typically done in current surgery, but rebuild new tissue using the material of the old.

Other surgery would benefit as well, or perhaps even more. In particular current surgical techniques disrupt cells, sever extracellular connective tissue, cut nerves and capillaries, and in general, do all sorts of damage that needs to be fixed by the healing process.

The surgeon with nanotech tools will not cut into you with a knife blade that is a slab of steel. He will use a machine, smaller than a current scalpel, that is more complex than an aircraft carrier. It will cut around cells rather than through them. It will send a detailed analysis of the tissue it is working on back to a database bigger than today's entire Internet. It will be able to terminate nerves and capillaries temporarily and flag them. When the operation is complete, other machinery will be able to reverse the separation, using the saved information and implanted flags, to a state that is nominal for healthy tissue. In other words, little or no healing will be necessary because the surgeon will put you back together, fixed and working, at the cellular level. There is no reason in principle that you could not have major surgery one day and play tennis, go dancing, or do a full day's work the next.

The same technology that will be used to implement the ultimate virtual reality, namely, tapping and spoofing sensory and motor nerve signals, would be invaluable as a replacement for anesthesia and its attendant dangers and recovery times. Not only would this speed the process and make it more comfortable, but you could be doing any thing you wanted, via *virtual reality* (VR) link, while it was happening. Suppose you'd rather be playing tennis than lying on the operating table. A robot your size and shape walks onto the court, and it has a data link to the anesthesia equipment. Signals from the robot are injected into your sensory nerves, so you see and hear what it does. Likewise, all the signals from your brain to your muscles are diverted and transmitted to the robot, so it actually does the actions you intend. The real sensory signals from your body, those of being operated on, are jammed, muted, or disconnected, and those from your brain similarly squelched, so that your body does not try a diving volley down the line just as the doctor is reconnecting your pancreas.

Spare Parts

Today, there is in some cases a choice between artificial and donated organs. With nanotechnology, virtually any organ will available in an artificial form, in general with better performance than the original.

Organs that perform physical functions, such as bones and muscles, are the easiest to replace. These are the ones today's technology handles best, like joints. Applications requiring power, like pumps such as an artificial heart, are complicated by methods of providing the power. This is ironic, since the cells of the body are crammed with mitochondria that turn sugar and fat into electricity—there are literally billions of highly efficient fuel cells at work inches from the prosthesis, but current medical technology can't take advantage of them.

Just like the rest of our technology, medicine is hampered by the gap between mechanical and molecular capabilities. We' are much better at breaking than fixing at the molecular scale. We could probably create a drug that would jam the ATP synthase motors in our mitochondria (it would be a deadly poison, of course), but we cannot use them to power our artificial hearts. Nanotechnology will close the gap. Nanotech organs could be powered by just the same fat- and sugar-burning reactions that cells use. Powered organs like heart and muscle will be straightforward. Organs that perform chemical functions, like the liver, and those that do physical sorting at the molecular level, like the kidneys and lungs, will be next.

Replacing organs that are body wide and connected to everything else, like the blood vessels, nervous system, or skin, will be the hardest of all. The problem is not so much the ability to make such an organ as the physical capability to replace it. Imagine you had to repair a faded tapestry by snipping out the faded thread, one strand at a time, and replacing it with new thread. You have your sewing scissors, plenty of thread, and a needle. Ready? The tapestry you have to fix covers the whole Empire State Building.

The same problem happens when we contemplate doing some thing to the body at the micron scale (much less the molecular scale!). It simply is not humanly possible to do that amount of work. The technology exists today to take a single cell and add or remove individual organelles under a high-powered microscope. If you could do that in a minute, working twenty-four hours a day, it would take you one hundred million years to do every cell in a human body.

Thus, once we figure out just what it is we want to do to cells, it is going to have to be done robotically. There just is not any chance of doing anything requiring human attention to any significant fraction of the cells, or even capillaries, in a body.

Single-cell Organ Replacement

Red blood cells carry oxygen by absorbing it into hemoglobin, a complex protein constructed to do so. However, we could carry far more oxygen in the same volume by pumping it into a pressure tank. With nanotechnology we can make the oxygen filter, pump, tank, some sensors, and control machinery, all smaller than a red blood cell. (Smaller because red cells fold to go through the smallest capillaries, but the tank would not.) A mechanism of this kind is called a respirocyte, invented by Robert Freitas. A respirocyte is essentially an artificial red blood cell.

The respirocyte might be about one cubic micron in volume, rounded smoothly to pass easily through capillaries, and shaped like any cylindrical pressure tank. The inside would have a partition that slides from one end to the other, separating volumes which contain oxygen and CO_2. It would have pumps and filters for oxygen, CO_2 and glucose, which it would use as a fuel. In terms of our five-million scale example, it's the size and complexity of a propane delivery truck.

The respirocyte would detect when it is in the lungs (by the high oxygen level) and pump in the oxygen and out the CO_2 it would do the opposite in the rest of the body. A respirocyte's tank might have a capacity of half a cubic micron, holding oxygen at 2,000 atmospheres pressure. (If the tank leaked, the resulting bubble would be just 10 by 10 by 10 microns, the size of a single ordinary cell, easily dissolved into the plasma and reabsorbed by other nearby respirocytes.) Replacing just a quarter of your red cells with respirocytes would give you storage for about 3 pounds of oxygen right there in your blood stream. For comparison, the body uses about a pound of oxygen a day at complete rest; or consumes it ten to twenty times as fast at the peak of exercise.

Besides letting you hold your breath all day, respirocytes would be good insurance against heart attacks. The oxygen in storage would sustain you for long enough to get help. The same is true if you were shot in the heart.

Another Freitas invention, the microbivore, replaces or augments white blood cells and antibodies. It is essentially a nano-Battlebot that physically destroys invading bacteria, viruses, and fungi. A shot of microbivores programmed to recognize a specific pathogen could cure a disease like measles or flu in minutes. People who live in the Northeast might get prophylactic doses for the Lyme disease spirochete, people in the tropics likewise for malaria.

Many other cells in the body could be replaced on a one-for-one basis. Some do simple functions such as pulling when told, that is, muscle cells. It would be perfectly feasible, for example, to supplement muscle with artificial winch cells. The trick would be putting them in place. For this purpose, replacement cells would have to be able to move within the tissues, find the right place to be, settle into place, and connect themselves to whatever was necessary. They could use a combination of internal recognition and external signals to locate themselves. If you did that, of course, you'd almost certainly want construction nanorobots to strengthen your bones and other ones to accelerate the maintenance on your joints.

Remember just how much machinery nanotechnology could put into the volume of an average human cell, a 10-micron cube. Scaled to the size where its parts are like car engine parts, that equivalent to a building seventeen stories high covering most of an acre. Or think of half a million car engine compartments. That much mechanism gives you the ability to do some very complex functions.

Note that the body contains signals—chemical and otherwise—for a cell to be able to tell what it should be doing. One of the most promising directions in current medical research involves stem cells. These are cells, like those in the fetus, that have not differentiated yet and thus still have the potential of becoming muscle, nerve, bone, fat, or whatever. Stem cells can regrow damaged organs." (It is the fetus connection, of course, that makes them controversial. However, the adult body also contains some stem cells, and there seem to be ways to convert other cells back into stem cells.) The point is that, although it is not completely understood yet, there's obviously enough information for the stem cell to "know" how to develop in its regrowth.

Cell Repair Machines

Ultimately, to attack things like cancer and aging, we are going to have to build robots that go into individual cells and fix genetic errors and repair other kinds of damage, such as those caused by free radicals and radiation. This is a bit trickier than replacing a cell. The size of a nanomachine that could move around inside a cell and do things is on the scale of one micron. This leaves it with five hundred car engines' worth of mechanism instead of half a million. It will obviously be possible to build machines that do particular functions, but a general-purpose repair robot that "understands" everything in the cell, and constantly checks and fixes it, is unlikely.

On the other hand, mechanisms that check for certain specific conditions seem quite feasible. One straightforward way of fighting cancer would be to place diagnostic units that detect when the cell's regulatory mechanisms have failed, and kill the cell (or block its reproduction). Cells already have such mechanisms. Half of the cur rent incidence of cancer is associated with a failure of the "p53 tumour suppressor" system, for example.' A nanomachine could either duplicate (or extend) the p53 system's function or check the p53-generating genes for mutations and correct them.

The p53 is not the only anticancer system in the cell—generally more than one thing has to break down before cancer develops. There are many things to check and fix, or to provide backups for. In the last-ditch case, nanomechanisms could be programmed to find cells exhibiting known *molecular markers* for cancer and destroy them.

Cancer is not the only untoward effect of nasty stray molecules, free radicals, and radiation. The molecular-scale damage these agents cause has various other bad effects as well. Again, these things are normally fixed by existing mechanisms in the cell or by letting a cell that is too badly messed up die, and others grow in its place.

In time, though, damage accumulates in spite of repair mechanisms. What is more, some of the cells' defenses play us false. For example, the Hayflick limit is a sort of counter that lets cells divide only so many times. It's one of the defenses against cancer. But after the human "design lifetime" of seventy years, the Hayflick limit begins to get in the way of the replacement of damaged cells with new ones.

Aubrey de Grey, a leading gerontologist, lists seven major problems at the cellular level that lead to senescence. They are:

1. Cells dying off (a normal process) and not being replaced.
2. Fat cells replacing working cells, and cells that have quit working but would not die on their own.
3. Mutations in the genes of a cell (its DNA) that make it quit working, produce things injurious to other cells, or become cancerous.
4. Mutations in the DNA of mitochondria. Mitochondrial DNA is much less well protected than the cell's main DNA, so it's more vulnerable. When this happens, the mitochondria don't produce the ATP the cell needs.
5. Garbage buildup inside cells. Cells constantly recycle the things they make inside, but every so often something gets made so wrongly that the normal mechanisms cannot disassemble it. Over the years, this junk builds up inside a cell and keeps it from working.
6. Garbage buildup outside cells. Again, the body has mechanisms for handling stuff like this, but there are a couple of cases where the buildup can defeat the body's cleanup crew. The main ones are the formations of plaques in the arteries causing atherosclerosis and the ones in the brain causing Alzheimer's disease.
7. Long-term chemical damage to the proteins outside the cell that have various physical functions, such as connective tissue, artery walls, the lens in the eye, and so forth."

These are arguably it. In the 1960s and 1970s gerontology regularly added new kinds of damage to the list as more of the mechanisms of aging were discovered. But the list has been stable for over twenty years. There is a reasonably good chance that once we learn to fix the damage these categories represent, we will be able to extend not only our lifetimes, but youthful vigour, to a significant extent.

It is important to point out that all these problems appear to be susceptible to some kind of treatment using nonnanotech techniques: stem cells fur number 1, targeted cell-killing drugs for 2, a whole raft of research aimed at cancer for 3, and so forth. These techniques are mostly still in the labs, but it's a very exciting time in gerontology research.

Like many of the other nanotech applications I have discussed here, an assault on aging is one that is right on the edge of current or near-future capabilities without nanotechnology. Thus we can have fairly strong confidence that we can do it with nanotechnology. Being able to measure, manipulate, and monitor at the molecular level can only improve our knowledge and capabilities.

Here is how we might expect nanotechnology to help address de Grey's seven pathologies:

1. *Cells dying off*: Stem cells are still the ticket here. However, a nanomachine may he able to prod a cell to divide where it would not have, by lengthening its telomeres, for example.
2. *Cellular that have quit or turn to fat*: Finding cells that are not working, or working well enough, is the hard part. "Census-taking" nanorobots that wander around checking cells might work. Killing the bad cells would be simple. There may even he some internal "switches" that can be thrown to change a fat cell back into muscle or whatever, or into a stem cell.
3. *Mutations in the genes*: Nanomachine to check or extend the cell's p53 system and other subsystems that repair DNA (or kill the cell if it is too damaged).
4. *Mutations in mitochondrial DNA*: Several options exist: check and fix mitochondrial DNA, manufacture the proteins it is supposed to specific or simply replace the mitochondria with mechanical ones.
5. *Garbage buildup inside cells*: Cell repair machines would have two basic options: break the junk down in place or carry it away. Which is best remains to be seen.
6. *Garbage buildup outside cells*: You don't even need cell repair machines for this, just larger ones that physically clean up the accumulations.

7. *Damage to extracellular structures*: Larger construction and repair nanorobots, as in point 6, but the machines check the material for damage and patch as needed.

Freitas has gone further than merely halting aging with the idea of "dechronification," that is, rolling back the clock. Dechronification will first arrest biological aging, then reduce your biological age by performing three kinds of procedures on each one of the tissue cells in your body:

First, a cell maintenance machine will be used on each cell to remove accumulating metabolic toxins and undegradable material. Afterward, these toxins will continue to reaccumulate slowly as they have all your life, so you will probably need a whole-body cleanout to prevent further aging, maybe once a year.

Second, chromosome replacement therapy can be used to correct accumulated genetic damage and mutations in every one of your cells. This might also be repeated annually, or less often.

Third, persistent cellular structural damage that the cell cannot repair by itself such as enlarged or disabled mitochondria can be reversed as required, on a cell-by-cell basis, using cellular repair devices.

The net effect of such interventions could be the continuing arrest of all biological aging, along with the reduction of current biological age to whatever new biological age is deemed desirable by the patient. These interventions may become commonplace several decades from today Notes Freitas: "If you are physiologically old and don't want to be, then for you, oldness and aging are a disease, and you deserve to be cured. Using annual checkups and cleanouts, and some occasional major repairs, your biological age could be restored once a year to the more or less constant physiological age that you select. I see little reason not to go for optimal youth—a rollback to the robust physiology of your early twenties would be easy to maintain and much more fun. That would push your Expected Age at Death up to around 700—900 calendar years. You might still eventually die of accidental causes, but you'll live ten times longer than you do now."

Ethical Issues

There seem to be two kinds of ethical issues facing medicine today with its new capabilities. The first is questions of whether we should actually do some cure or improvement, for its own sake. The second is the question of making humans, or parts of them, specifically to cure other people. I will leave questions of the first kind for the chapter on transhumanism, although as far as cures are concerned, the issues seem at least somewhat more straightforward. For the second kind, though, nanotechnology may be able to finesse some of the thornier problems.

The reason is that the nanotech route to certain kinds of capability involves building machines that are clearly machines and have specific, planned functions. The biotechnology route, however, leaves you in the gray area between handling parts and handling people. Questions like the morality of cloning, or buying and selling organs, tend to arise. With nanotechnology, where you can fix the patient's own cells and build purely mechanical parts as good as the originals, those particular questions don't stand in the way of healing.

Live Long and Prosper

An American man born in 1850 had a life expectancy of about thirty- eight years. An American man born in 1990 has a life expectancy of about seventy-three. In other words, for every four years that passed during the period, life expectancy increased a year. (It was not steady, of course—wars and medical advances like penicillin varied the rate.) Besides medicine, much of the improvement was due to the effects of the first industrial revolution: proper food, clothing, shelter, sanitary facilities, and so forth.

The demographic human lifespan is beginning to bump up against old age, as productivity and medicine have removed the bulk of the causes that used to cut it short. But we have seen aging is not mysterious; it is caused by understandable phenomena that we will soon have the means to combat. There is no reason, then, to expect the historical trend to stop. More specifically, if you can hang around for the next few decades, you can probably expect to be here for quite a while longer than that.

In 1960 total private and public spending on health was $27 billion. In 2000 the figure was $1.3 trillion, nearly fifty times as much. By 2010 it is expected to double again to $2.6 trillion. The obvious question to anyone faced with today's high health insurance costs is whether nanotechnology will be able to apply the same kind of drastic cost reductions to medicine that we have projected for manufacturing and other high-tech applications. In the long run quite likely. In the short run, not so likely. There are many structural reasons for this, but the basic cause will be that every advance nanotechnology brings will be swallowed up in the great struggle to understand and then defeat cancer, AIDS, and old age—not to mention heart disease, arthritis, and the common cold. It may be late in the twenty-first century and technology enormously advanced on other fronts with capabilities we can now only dimly dream of, before these are conquered. After that, the cost reductions will begin to take hold and finally settle down to reasonable maintenance costs for the human machine. After that we'll face the real problem: what to give on a 250th or 500th anniversary.

9

Nanotechnology Revolution

The English chemist John Dalton first proposed the scientific theory of the atom two hundred years ago. Since then we have seen chemists come to understand the elements and their interactions, we have seen engineers make and use new materials to improve our lives, we have seen physicists demonstrate that even atoms are divisible, and we have seen warriors unleash the power of the atomic nucleus. In these two centuries we have amassed an enormous understanding of—and wielded an increasing control over—the fundamental units of matter.

Today, in the young field of nanotechnology, scientists and engineers are taking control of atoms and molecules individually, manipulating them and putting them to use with an extraordinary degree of precision. Word of the promise of nanotechnology is spreading rapidly, and the air is thick with news of nanotech breakthroughs. Governments and businesses are investing billions of dollars in nanotechnology R&D, and political alliances and battle lines are starting to form. Public awareness of nanotech is clearly on the rise, too, partly because references to it are becoming more common in popular culture—with mentions in movies (like *The Hulk* and *The Tuxedo*), books (including last year's Michael Crichton bestseller, *Prey*), video games (such as the "Metal Gear Solid" series), and television (most notably in various incarnations of Star Trek).

Yet there remains a great deal of confusion about just what nanotechnology is, both among the ordinary people whose lives will be changed by the new science, and among the policymakers who wittingly or unwittingly will help steer its course. Unsurprisingly, some of the confusion is actually caused by the increased attention—sensationalistic reporting and creative license have done little to prepare society for the hard decisions that the development of nanotechnology will make necessary.

Much of the confusion, however, comes from the scientists and engineers themselves, because they apply the name "nanotechnology" to two different things—that is, to two distinct but related fields of research, one with the potential to improve today's world, the other with the potential to utterly remake or even destroy it. The meaning that nanotechnology holds for our future depends on which definition of the word "nanotechnology" pans out. Thus any understanding of the implications of nanotechnology must begin by sorting out its history and its strange dual meaning.

Mainstream Nanotechnology

Nanotechnology got going in the second half of the twentieth century, although a few scientists had done related work earlier. For instance, as part of an 1871 thought experiment, the Scottish physicist James Clerk Maxwell imagined extremely tiny "demons" that could redirect atoms one at a time. And M.I.T. professor Arthur Robert von Hippel (born in 1898 and still alive today) became interested in molecular design as early as the 1930s; he coined the term "molecular engineering" in the 1960s.

Usually, though, the credit for inspiring nanotechnology goes to a lecture by Richard Phillips Feynman, a brilliant Caltech physicist who later won a Nobel Prize for "fundamental work in quantum electrodynamics." He is best remembered today for his clear and quirky classroom lectures and for his critical role on the presidential commission that investigated the *Challenger* accident. On the evening of December 29, 1959, Feynman delivered an after-dinner lecture at the annual meeting of the American Physical Society; in that talk, called "There's Plenty of Room at the Bottom," Feynman proposed work in a field "in which little has been done, but in which an enormous amount can be done in principle."

"What I want to talk about," Feynman said, "is the problem of manipulating and controlling things on a small scale. As soon as I mention this, people tell me about miniaturization, and how far it has progressed today ... But that's nothing; that's the most primitive, halting step in the direction I intend to discuss."

Feynman described how the entire *Encyclopaedia Britannica* could be written on the head of a pin, and how all the world's books could fit in a pamphlet. Such remarkable reductions could be done as "a simple reproduction of the original pictures, engravings, and everything else on a small scale without loss of resolution." Yet it was possible to get smaller still: if you converted all the world's books into an efficient computer code instead of just reduced pictures, you could store "all the information that man has carefully accumulated in all the books in the world ... in a cube of material one two-hundredth of an inch wide—which is the barest piece of dust that can be made out by the human eye. So there is *plenty* of room at the bottom! Don't tell me about microfilm!" He boldly declared that "the principles of physics, as far as I can see, do not speak against the possibility of maneuvering things atom by atom"—in fact, Feynman saw atomic manipulation as inevitable, "a development which I think cannot be avoided."

In his lecture, Feynman pointed out several avenues for research that would later come to define nanotechnology, such as making computers much smaller and therefore faster, and making "mechanical surgeons" that could travel to trouble spots inside the body. Feynman admitted that he didn't have a clear conception of how such tiny machines might be used or created, but to help get things going, he offered two prizes: $1,000 to the first person to make a working electric motor that was no bigger than one sixty-fourth of an inch on any side, and another $1,000 to the first person to shrink a page of text to 1/25,000 its size— the dimension necessary to fit the *Encyclopaedia Britannica* on the head of a pin. (He awarded the former prize in 1960, the latter in 1985.)

Although Feynman's lecture is, in retrospect, remembered as a major event, it didn't make much of a splash in the world of science at the time. Research in the direction he suggested didn't begin immediately, and nanotechnology was slow to take off. Feynman himself didn't use the word "nanotechnology" in his lecture; in fact, the word didn't exist until 15 years later, when Norio Taniguchi of the Tokyo University of Science suggested it to describe technology that strives for precision at the level of about one nanometer.

A nanometer is one billionth of a meter. The prefix "nano-" comes from the Greek word *nanos*, meaning dwarf. (Scientists originally used the prefix just to indicate "very small," as in "nanoplankton," but it now means one-billionth, just as "milli-" means one-thousandth, and "micro-" means one-millionth.) If a nanometer were somehow magnified to appear as long as the nose on your face, then a red blood cell would appear the size of the Empire State Building, a human hair would be about two or three miles wide, one of your fingers would span the continental United States, and a normal person would be about as tall as six or seven planet Earths piled atop one another.

In 1981, scientists gained a sophisticated new tool powerful enough to allow them to see single atoms with unprecedented clarity. This device, the scanning tunneling microscope, uses a tiny electric current and a very fine needle to detect the height of individual atoms. The images taken with these microscopes look like tumulose alien landscapes—and researchers learned how to rearrange those landscapes, once they discovered that the scanning tunneling microscope could also be used to pick up, move, and precisely place atoms, one at a time. The first dramatic demonstration of this power came in 1990 when a team of IBM physicists revealed that they had, the year before, spelled out the letters "IBM" using 35 *individual atoms* of xenon. In 1991, the same research team built an "atomic switch," an important step in the development of nanoscale computing.

Another breakthrough came with the discovery of new shapes for molecules of carbon, the quintessential element of life. In 1985, researchers reported the discovery of the "buckyball," a lovely round molecule consisting of 60 carbon atoms. This led in turn to the 1991 discovery of a related molecular shape known as the "carbon nanotube"; these nanotubes are about 100 times stronger than steel but just a sixth of the weight, and they have unusual heat and conductivity characteristics that guarantee they will be important to high technology in the coming years.

But these exciting discoveries are the exception rather than the rule: most of what passes for nanotechnology nowadays is really just materials science. Mainstream nanotechnology, as practiced by hundreds of companies, is merely the intellectual offspring of conventional chemical engineering and our new nanoscale powers. The basis of most research in mainstream nanotech is the fact that some materials have peculiar or useful properties when pulverized into nanoscale particles or otherwise rearranged.

Seen this way, mainstream nanotechnology isn't truly new; we've been unwitting nanotechnologists for centuries. One official from the National Science Foundation told Congress that photography, of all things, is a subset of nanotechnology —and a "relatively old" one at that! But if the term "nanotechnology" is to be used that loosely, why not reach much further back into history? Renaissance artists used paints and glazes that got their appealing colour and iridescence from nanoparticles. The ancients, too, found uses for nanoparticles of soot. On and on it goes, back through the ages.

A great many of today's mainstream nanotechnologists are simply following in that tradition, using modern techniques to make tiny particles and then finding uses for them. Among the products that now incorporate nanoparticles are: some new paints and sunscreens, certain lines of stain- and water-repellent clothing, a few kinds of anti-reflective and anti-fogging glass, and some tennis equipment. Cosmetics companies are starting to use nanoparticles in their products, and pharmaceuticals companies are researching ways to improve drug delivery through nanotech. Within a few years, nanotechnology will most likely be available in self-cleaning windows and flat-screen TVs. Improvements in computing, energy, and medical diagnosis and treatment are likely as well.

In short, mainstream nanotechnology is an interesting field, with some impressive possibilities for improving our lives with better materials and tools. But that's just half the story: there's another side to nanotechnology, one that promises much more extreme, and perhaps dangerous, changes.

Molecular Manufacturing

This more radical form of nanotechnology originated in the mind of an M.I.T. undergraduate in the mid-1970s. Kim Eric Drexler was specializing in theories of space travel and space colonization in college when he first thought of using DNA to make computers. But why stop there? He soon realized that the biological "machinery" already responsible for the full diversity of life on Earth could be adapted to build nonliving products upon command. Molecule-sized machines, originally derived from those found in nature, could be used to manufacture just about anything man wished. *Anything*.

Drexler, who began to develop these theories before he'd heard of Feynman's lecture, first published his ideas in a 1981 journal article. Five years later, he brought the notion of molecular manufacturing to the general public with his book *Engines of Creation*. An astonishingly original work of futurism, *Engines* presented Drexler's nanotech theories and pointed out how nanotechnology would revolutionize other areas of science and technology—leading to breakthroughs in medicine, artificial intelligence, and the conquest of space.

At the heart of Drexler's vision for molecular manufacturing was a kind of nanomachine called an "assembler," which can "place atoms in almost any reasonable arrangement," thus allowing us to "build almost anything that the laws of nature allow to exist." It would take millions and millions of assemblers to make a product big enough for us to use—so in order for molecular manufacturing to work, assemblers must be capable of replicating themselves; as each "generation" of assemblers replicated itself, the overall number of assemblers would grow exponentially.

In one of the most striking passages of Drexler's book, he describes how molecular manufacturing could be used to build—to *grow*, really—a large rocket engine. Replicating assemblers would be pumped into a vat, and all the plans for the rocket engine would be stored on a single "seed." With the addition of fuel for the assemblers and raw materials for the construction, the engine would be completed in "less than a day" and would require "almost no human attention." The final product would be "a seamless thing, gemlike," light and strong— instead of a clunky, "massive piece of welded and bolted metal." If you wanted, you could "exploit nanotechnology more deeply" by building engines that repair themselves, or that "take different shapes under different operating conditions."

Drexler further imagined how nanotechnology could completely reshape everyday life. "It should be no great trick, for example, to make everything from dishes to carpets self-cleaning, and household air permanently fresh." Fresh food—"genuine meat, grain, vegetables, and so forth"—could be produced in the home. Suits made with nanotechnology could be used for virtual reality, simulating "most of the sights and sensations of an entire environment." And nanotechnology could make "some form" of telepathy "as possible as telephony."

Such powers seem like magic, a comparison that Drexler acknowledged: nanotechnology, he wrote, could make possible a device that "might aptly be called a 'genie machine.'" In a later book, he described in general terms how such a machine might be designed. Other writers, following in Drexler's footsteps, have imagined other grant-any-wish tools—like "utility fog," a theoretical swarm of tiny robots that could "simulate the physical existence of almost any object" and can thus "act as shelter, clothing, telephone, computer, and automobile." As envisioned by John Storrs Hall, the techie who dreamed the stuff up, the utility fog "that was your clothing becomes your bath water and then your bed."

Clearly, the Drexlerian notion of nanotechnology differs vastly from the nanotech products of today. Compare, for instance, how the two divergent visions of nanotechnology would differently affect one small aspect of human life: cosmetics. Mainstream nanotechnology will soon be used by cosmetics companies to help their current products—makeup, lotions, sunscreen, and so forth—last longer and work better. But if Drexler's version of nanotechnology were to come to fruition, the beauty industry would be revolutionized: nanomachines could precisely adjust your hair and skin colour to your liking; wrinkles could be smoothed and excess fat removed; one writer suggests it would even become possible to mold the face and body to whatever shape might be desired. Each person who cared to could achieve his or her own ideal of physical perfection or, for that matter, whatever frightening or gruesome effect they wanted. Many who never liked their own youthful appearance will opt instead to copy some popular model or other sex symbol. It could become very confusing, with dozens of pop-idol look-alikes crowding the parks and boulevards of our future metropolis. Some may not relish the prospect, but we may never see the last of the Elvis clones.

So while mainstream nanotech gives you better eyeshadow, Drexler's nanotech gives you a whole new face—yet these two technologies of profoundly different potential share one name. "If research on waterproof fabric coatings is 'nanotechnology,' then the term has become almost meaningless," Drexler told *Wired News* in June. Drexler himself now talks about his kind of nanotech as "molecular nanotechnology" and "molecular manufacturing." Other names have been suggested, too: one observer has argued that Drexler should start using the ugly word "mechutechnology." But for most people, one umbrella term describes both the mainstream approach and Drexler's more radical vision.

But is Drexler's nanotechnology realistically possible? Will we truly be able to watch houses build themselves from the ground up, to transform garbage into steak, to populate the world with Elvis look-alikes? Drexler's book *Engines of Creation* is an extraordinary exercise in prolepsis: he meticulously refutes every technical objection he can anticipate. Will thermal vibrations make his molecular machines impossible? (No.) What about radiation? (No.) Quantum uncertainty? (No!) To shore up his technical arguments for the feasibility of his vision, he further expanded on his ideas in the world's first nanotechnology textbook. *Nanosystems* (1992), a dense volume that grew out of a class he taught at Stanford, is crammed with equations and diagrams and designs for molecular machines, and it has gone far to put Drexler's nanotechnology on sound technical footing.

To date, no scientist or engineer has been able to make a rock-solid argument showing the impossibility of molecular manufacturing as Drexler envisions it. A few critics have challenged Drexler on technical points, most prominently Richard Errett Smalley, the Rice University chemist who won a Nobel Prize for discovering the new class of carbon molecules that includes buckyballs and carbon nanotubes. In 1999, in written testimony to a congressional subcommittee, Smalley claimed that Drexler's version of nanotechnology is "just a dream" and "will always remain a fantasy" because "there are simple facts of nature that prevent it from ever becoming a reality."

When these claims were repeated in a 2001 article in *Scientific American* and again in public this year, Drexler and his allies responded with strongly worded public letters, accusing Smalley of basing his challenge on a straw man. Without getting into the technical details of the dispute, the essence of Drexler's response is devastating: If "atomically precise structures" are "fundamentally unfeasible, then so is life" itself, he wrote. Since enzymes and ribosomes and other molecular "machines" work in nature, man-made molecular machines should work, too.

Drexler and his supporters have made short shrift of other critics as well. Yet if no one has mounted a serious and sustained challenge to demonstrate a fatal flaw, or even a major error, in Drexler's vision of nanotechnology, why is it so often disparaged by those involved in mainstream nanotech? Molecular manufacturing has been called pseudoscience, science fiction, and unrealistic utopianism, and Eric Drexler himself has suffered repeated ad hominem attacks.

One reason for the animosity of the mainstreamers is their fear that Drexler's talk of the great boon and bane of nanotechnology will cast a pall over their own modest research—giving nanotech a reputation for being fantastical or hazardous.

A second explanation for why mainstream nanotech experts pooh-pooh Drexler is that they simply don't know what they're talking about. Drexler's kind of nanotechnology is so newfangled that it doesn't fit neatly into any single division of modern science or technology. (That fact actually caused difficulties when Drexler tried to obtain his Ph.D. at M.I.T.; he was eventually awarded an interdisciplinary degree—the world's first doctorate in nanotechnology.) It's difficult to find "appropriate critiques of nanotechnology designs," Drexler wrote in a 2001 article in *Scientific American*, since "many researchers whose work seems relevant are actually the wrong experts—they are excellent in their discipline but have little expertise in systems engineering. The shortage of molecular systems engineers will probably be a limiting factor in the speed with which nanotechnology can be developed."

No doubt some of the criticism of Drexler's nanotechnology is rooted in this important fact: nobody knows how to make the key component of his molecular manufacturing system, the assembler. Although Drexler and his supporters have come up with lots of designs for molecular machines and plans for how they would function, there still isn't any way to make them real. When someone figures out how to make the miniscule workhorses of molecular manufacturing—the critical moment of discovery that Drexler calls "the assembler breakthrough" —the rest may quickly fall into place, and the world could be transformed abruptly and forever.

Nanomedicine

Some people find Eric Drexler's vision of the nanotech future so compelling that they embrace it with religious fervor. This is not a new observation; a 1989 *Economist* article about Drexler spoke of his "gospel of nanotechnology." The 1995 book *Nano* by Ed Regis includes an entire chapter called "Brother Eric's Nanotech Revival," describing the sense of awe that Drexler's lectures would inspire in members of the audience: "There was a veiled feeling of being one of the Elect, the Select, the Knowledgeable, the Chosen."

A half-century ago, philosopher and technology critic Jacques Ellul argued that the rise of technology leads to the decline of traditional spirituality, as man transfers "his sense of the sacred ... to technique itself." We develop a "worship of technique," Ellul said, and we associate our technology with a "feeling of the sacred." Drexler's nanotechnology is perfectly suited to arouse religious enthusiasm. It involves incredible, invisible powers. The all-important "assembler breakthrough" is akin to a Second Coming or a Judgment Day. And there's even an afterlife: cryonics.

Nanotechnology is especially appealing to those in the growing ranks of what Charles T. Rubin, in the previous issue of this journal, usefully dubbed the "extinctionist project": the transhumanists, posthumanists, extropians, and others who seek to completely remake human nature. Nanotechnology is central to their vision of a future of agelessness, immortality, and rebirth.

They place their hopes in nanomedicine, a field that would repair or improve the body from the inside out, with a precision and delicacy far greater than that of the finest surgical instruments available today.

Science fiction envisioned tiny internal medical procedures long ago; in the 1966 movie *Fantastic Voyage*, a medical staff boards an experimental submarine which is then drastically miniaturized and injected into a patient in order to destroy a deadly blood clot. Of course, miniaturizing humans is preposterous, but in the 1960s it was hard to imagine any other way to make tiny machines *intelligent* enough to reach and repair damage inside the body. But nanomachines are certainly small enough, and with programmed instructions, they can be smart, too.

The world's leading expert on nanomedicine is Robert A. Freitas, Jr., a polymath with a law degree who worked on numerous space-related projects before becoming involved in nanotechnology. Currently employed as a researcher at the nanotech firm Zyvex, he is one of only a handful of people who can claim to have made major theoretical contributions to Drexlerian nanotechnology.

Freitas is currently several years into the writing of an exhaustive four-volume series called *Nanomedicine*, the first technical work on the subject. In the first massive volume (published in 1999), he offers technical speculations on how nanorobots might navigate, sense their surroundings, and move through the body; how they might detect problems and communicate with one another; and how they might change shape and obtain energy. The second volume of *Nanomedicine*, due out this year, will examine "biocompatibility"—how nanorobots might interact with the body, especially the immune system.

Many of the tools of nanomedicine could be used for either therapy or enhancement. Take, for example, the "respirocyte," an artificial red blood cell about which Freitas has theorized. Respirocytes, capable of delivering oxygen hundreds of times more efficiently than real red blood cells, would be invaluable in the treatment of various respiratory and cardiovascular disorders, or as a substitute for real blood during transfusions. But they would also have "a variety of sports, veterinary, battlefield and other applications"; they could be used to boost a mountain climber's endurance, to help a diver hold his breath for hours, or to enable a soldier to fight harder.

And the respirocyte is among the simplest medical nanomachines imaginable. Others might be able to repair cells and fix damaged DNA; to remove toxins, clean out cholesterol, and eliminate scar tissue; to destroy cancer cells and fight countless diseases. And the same nanotechnology that keeps your body healthy can indefinitely stave off senescence. The process of aging, Drexler argued in *Engines of*

Creation, is "fundamentally no different from any other physical disorder," so cell repairing nanomachines should, in theory, be able to halt aging or reverse it. You can pick the age you want to be—in fact, you can play mix and match: give yourself the distinguished hairline of a fifty-year-old, the sturdy frame of a thirty-year-old, the lusty libido of a twenty-year-old, and the keen eyesight of a ten-year-old.

Even the Grim Reaper is in for tough times: Death may already be "slave to Fate, Chance, kings, and desperate men," but in the age of nanotechnology, Death will increasingly obey the whims of Tom, Dick, and Harry, too. Molecular machines will bridge the gap between living matter and nonliving matter, making the border between life and death much fuzzier. In the age of nanotechnology, a person might intentionally put himself into stasis, perhaps to "time travel" dreamlessly into the future, or to wait out a centuries-long interstellar voyage. Even today, hundreds of people of sufficient means are making plans to freeze themselves in hopes that nanotech will someday restore them; these people are willing to shell out big bucks to cryonics companies that promise to preserve their corpses, or some meaningful fraction thereof, until the prospect of reanimation becomes realistic.

There are, however, some foreseeable limits to nanomedicine. While nanomachines might one day be able to restore and maintain the body, there is no guarantee that they'll be able to keep the *mind* intact. Some brain damage can be physically fixed, but lost memories and personality—the brain's software (figuratively speaking)—will be irretrievable. If bits of your mind are lost, "repair machines could no more restore them than art conservators could restore a tapestry from stirred ash," as Drexler has said. But of course many of those engaged in the extinctionist project have a solution in the works: they seek to reduce the mind to software (literally) so the contents of your brain can be as downloadable and fungible tomorrow as digital video and music are today.

How Soon?

Estimates on how long we have to wait for major breakthroughs in nanotechnology vary greatly. Robert Freitas told one interviewer that the kind of nanomedicine he envisions is "at least 10 to 20 years away"; in a different interview he put the number at 40 years. Another nanotech expert says molecular manufacturing is 20 or 30 years away. We'll have to wait at least ten years before we can ride in "superintelligent" airplanes enhanced with nanotechnology, according to a Boeing executive. An all-purpose nanotech entertainment system could "arrive on the scene around the year 2020," according to one writer. The British Ministry of Defense says nanotech won't hit its stride any earlier than 20 or 30 years from now, but a Canadian expert says it will start to dramatically change our lives in the next 10 to 20 years. Ray Kurzweil, the technologist, predicts in his book *The Age of Spiritual Machines* that nanotech will be used in manufacturing by 2019—and that by 2049, smart swarms and nanotech food will be feasible. The U.S. government projects that the worldwide nanotechnology market will exceed $1 trillion by 2015, although one group opposed to nanotechnology puts it more ominously: by 2015, the controllers of nanotechnology "will be the ruling force in the world economy."

While there have been a few indications of progress in nanotechnology in the past two or three years, the present booming interest in all things "nano" is bound to quicken the pace of discovery. In the U.S., so many states are subsidizing nanotech research that a *New York Times* reporter whose job was to read governors' "State of the State" speeches in 2001 found herself asking: "Are there enough nanotechnological researchers to go around?" Governments in Europe and Asia are also putting money into nanotech, including Switzerland, Germany, Britain, China—and even Iran.

Businesses around the world are spending heavily on mainstream nanotech, pouring more than $3 billion into nanotech R&D this year alone, according to one estimate. A recent survey showed that "13 of the top 30 Dow component companies discuss nanotechnology on their websites." But all the nanotech buzz is destined to attract con artists and frauds, too. Some companies doing work completely unrelated to nanotechnology have incorporated "nano" into their names, in hopes of getting money from gullible investors caught up in the hype. And earlier this year, a major conference on nanotech was canceled under mysterious circumstances; it now appears that the whole thing was a scam to get money from the attendees.

If anything is likely to dampen the nanotech boom, it is the prospect of regulation. In the past year, mainstream nanotech has suddenly come under scrutiny from researchers and activists worried that nanoparticles could endanger public health or harm the environment. So far, there has been very little precautionary research on the safety of nanoparticles; indeed, when Rice University's Center for Biological and Environmental Nanotechnology conducted a survey of the scientific literature relating to nanoparticles—"a field with more than 12,000 citations a year"—they found *no* documented research on the risks of nanoparticles. Vicki L. Colvin, the Center's director, told Congress last April that the safety of nanoparticles should be determined through immediate and thorough tests. "From asbestos to DDT we have, as a society, paid an enormous price for not evaluating toxicological and ecosystem impacts before industries develop," she said. Her organization has started investigating nanoparticle safety, and in July the main U.S. lobbying group for the nanotechnology industry (the NanoBusiness Alliance) announced that it, too, would start studying the health and environmental safety of nanoparticles. Similar inquiries have begun in Britain and the European Union.

Beyond the Gray Goo

The health and environmental threats posed by mainstream nanotech are far less frightening than the hypothetical dangers of the Drexlerian flavor of nanotech. In an infamous article in *Wired* magazine in 2000, technologist Bill Joy made the case for halting nano-research because of the possibility that we might wipe out all life on Earth. Joy's article stirred up a hornet's nest of controversy, even though the idea had been around for a long time: the apocalypse he described was based on a theory that had first been suggested more than a decade earlier in *Engines of Creation*.

For molecular manufacturing to work, Drexler's assemblers would have to replicate themselves, just as tiny organisms make duplicates of themselves. But what if something went wrong—what if the replication spiraled out of control? Speed-breeding assemblers could devour all life on Earth in short order. According to *Engines of Creation*, "among the cognoscenti of nanotechnology"—presumably meaning the author and his friends—"this threat has become known as the 'gray goo problem'":

Though masses of uncontrolled replicators need not be gray or gooey, the term "gray goo" emphasizes that replicators able to obliterate life might be less inspiring than a single species of crabgrass. They might be "superior" in an evolutionary sense ... The gray goo threat makes one thing perfectly clear: we cannot afford certain kinds of accidents with replicating assemblers. Gray goo would surely be a depressing ending to our human adventure on Earth, far worse than mere fire or ice, and one that could stem from a simple laboratory accident ... We must not let a single replicating assembler of the wrong kind be loosed on an unprepared world.

In time, Drexler backed away from the gray goo scenario, reasoning that no one would design a self-replicating assembler capable of surviving in nature. "Consider cars," he wrote in 1990. "To work, they require gasoline, oil, brake fluid, and so forth. No mere accident could enable a car to forage in the wild and refuel from tree sap ... It would be likewise with simple replicators designed to work in vats of assembler fluid"—no right-minded engineer would create replicators that could exist in the wild.

But a terrorist might. Or an enemy nation. Biosphere-destroying self-replicators may not arise as the result of an inadvertent scientific slip-up, but they might be designed intentionally by those seeking to bring destruction or wreak havoc.

In the wake of Bill Joy's screed and the gray goo frenzy it inspired, Robert Freitas, the nanomedicine expert, wrote the first serious, technical analysis of the gray goo scenario. His paper—to which he gave the whimsical title "Some Limits to Global Ecophagy by Biovorous Nanoreplicators, with Public Policy Recommendations"—made estimates and calculations relating to the speed of replication and the rate of dispersal of self-replicators in the wild. If certain unlikely conditions are arranged just right, it is theoretically possible, Freitas found, for self-replicators to destroy the planet's entire biosphere in under three hours—but such a high-speed attack would instantly cause a massive spike in temperature, alerting authorities to the situation and allowing them to respond. Conversely, it is theoretically possible for biosphere-eating self-replicators to create an almost undetectably small increase in temperature—but then it would take them twenty months to complete their task, leaving plenty of time to observe the destruction and organize a defense.

Instead of going through all the trouble of designing self-replicators that could eat everything, why not design some that could make a more focused attack? "The classic example," Freitas says, "is tire rubber and asphalt tar binder; cars, trucks and airplanes roll on roads and tarmacs worldwide." Imagine the damage to the global economy if all the world's roads became soup overnight. "Other vectors with similar properties include cotton, polyester or other uniform textiles, insulation on electrical wiring, and paper money."

And though Freitas doesn't say as much, it seems possible that self-replicating nanobots could be designed to target and destroy a specific species. Perhaps they could be tailored to attack only humans—or just specific groups of humans, or just a specific individual.

Nanoweapons could be smarter than conventional biological weapons, with a more precise lethality and potential to cause diseases unlike any seen before. What's more, nanotechnology could also be used to aid in the manufacture and targeting of conventional bioweapons. (There has also been some recent speculation that nanotechnology could be combined with nuclear weaponry. Analysts at the Acronym Institute for Disarmament Diplomacy and *Jane's Chem-Bio Web* have theorized that nanotechnology could play a part in the creation of so-called "fourth-generation" nukes with small, lowyield warheads. But they apparently confuse nanotechnology with microtechnology, and they make confusing and contradictory assumptions about how the technologies needed to enable fourth-generation nukes will evolve.)

Aside from nanotech's potential as a weapon of mass destruction, it could also make possible totally novel forms of violence and oppression. Nanotechnology could theoretically be used to make mind-control systems, invisible and mobile eavesdropping devices, or unimaginably horrific tools of torture. Yes, it's true that defensive applications of nanotechnology would develop alongside offensive ones, but that hardly mitigates the potential for enormities and catastrophes.

To save the world, Bill Joy argues that we must relinquish our pursuit of nanotech knowledge. He recommends international treaties and a verification regime. But that would be exceedingly difficult, since nanotechnology isn't just a science of the small, it's also a *small science*: it doesn't require giant equipment or big laboratories or gigantic budgets, and most of the work is conducted in small labs distributed around the world rather than in a few centralized behemoth facilities. Scientists wishing to hide their nanotech research programs could easily disguise them as other projects in chemistry or physics. The allure of nanotechnology is so great that relinquishment could only work if it were enforced through "detailed, universal policing on a totalitarian scale," as Eric Drexler has worried, or if some horrible nanotech-related disaster shocked the world into giving up on nano.

No callow cheerleader for the nanotech revolution, Drexler in 1986 cofounded the Foresight Institute, a California-based nonprofit organization "formed to help prepare society for anticipated advanced technologies," especially nanotechnology. Among the Institute's projects is a set of proposed "Foresight Guidelines on Molecular Nanotechnology," first drafted in 1999. The guidelines forbid the creation of nanobots capable of "replication in a natural, uncontrolled environment," and provide several other principles for nanotechnologists.

According to the Foresight website, the Guidelines "might eventually be enforced via a variety of means, possibly including lab certifications, randomized open inspections, professional society guidelines and peer pressure, insurance requirements and policies, stiff legal and economic penalties for violations, and other sanctions."

Politics of Nanotech

So far, no nanotech businesses have adopted the Foresight Guidelines—after all, most firms are working on mainstream nanotech, not the riskier kind. Besides, nanotech companies have no motivation to regulate themselves, since it seems unlikely that they will be regulated by government any time soon. But this may change as the politics of nanotechnology begin to take shape.

Some agencies in the federal government have been involved in nanotechnology since at least the early 1980s, most notably the U.S. Naval Research Laboratory. By 1997, the federal government was annually investing $116 million in nanotech; that figure had doubled by 1999.

In 2000, the Clinton Administration pushed for more subsidies for nanotech and the creation of a National Nanotechnology Initiative (NNI) that would coordinate the nanotech work of six different agencies. President Clinton alluded to nanotechnology in that January's State of the Union Address, when he spoke of "materials ten times stronger than steel at a fraction of the weight, and—this is unbelievable to me—molecular computers the size of a teardrop with the power of today's fastest supercomputers." His administration worked hard to sell the proposal to Congress; as one official from the Clinton White House told *Scientific American*, "You need to come up with new, exciting, cutting-edge, at-the-frontier things in order to convince the budget- and policy-making apparatus to give you more money."

Congress couldn't resist, and the NNI was approved with an initial budget of $422 million. President Bush, in the first year of his administration, asked for another hundred million dollars for nanotech, and added another handful of agencies to the NNI. Bush's budget proposals for FY2003 and FY2004 further boosted the nanotech budget—despite the flagging economy and the war on terrorism. (In fact, some NNI proponents have used the war on terrorism to make the case for increasing nanotech funding; they say nanotech research can help build tools to detect weapons of mass destruction.)

Flush with nanotech cash, the National Science Foundation recently started a program to teach high school and elementary school students about nanotechnology, "with introduction to preliminary concepts as early as kindergarten," according to the *Christian Science Monitor*. "Business, industry, and higher-education leaders agree, saying early education gives students a jump on a job market many expect to blossom in the future."

Perhaps the most prominent federal entity under the NNI umbrella is the Department of Defense, which in May unveiled its new $50 million Institute for Soldier Nanotechnologies at M.I.T. The Institute, which treats soldiers as "integrated platform systems" rather than human beings, will bring together M.I.T. scientists, military officers, and researchers from private industry to develop lighter, stronger clothes and equipment for the Army. Some of the projects being suggested include an "exoskeleton" or "dynamic armor," which could become hard or soft at a soldier's command, and other clothes that could store energy— like the energy wasted in every footstep—and employ it later to give the soldier superhuman strength. All the technologies being developed at the Institute—like all other nanotech projects publicly acknowledged by the Defense Department—are essentially defensive, not offensive, in nature, so they are unlikely to incite opposition.

At the same time, because the benefits of nanotechnology are still largely uncertain, there is not yet a natural constituency for nanotech legislation— except for the nanotech companies themselves. They are represented by the New York-based NanoBusiness Alliance, a trade group founded in 2001 by F. Mark Modzelewski, who acts as the Alliance's executive director. Modzelewski, who modeled his group after the Biotechnology Industry Organization, was a lowranking official in the Clinton Administration—which hasn't stopped him from making Newt Gingrich, that starry-eyed technophile, the Alliance's honorary chairman. Gingrich told the *Forbes/Wolfe Nanotech Report* that he believes that "those countries that master the process of nanoscale manufacturing and engineering will have a huge job boom over the next twenty years, just like aviation and computing companies in the last forty years, and just as railroad, steam engine and textile companies were decisive in the nineteenth century."

Since the politics of nanotechnology are still immature, there is no prominent opponent of nanotechnology in the nation's capital or even a unifying rationale for such opposition. The most organized opposition to nanotechnology has come from the ETC Group, a liberal Canadian environmental outfit that has published a series of harshly critical reports on nanotechnology—some of them detailed and provocative. In late July, Greenpeace issued its first report on nanotechnology, with ambiguous conclusions. A few other environmentalist groups have spoken out against nanotechnology, but there hasn't yet been any movement comparable to the massive international campaigns against genetically modified foods. It is safe to speculate that these leftist groups will in time coalesce into an anti-nanotech front, using the rhetoric of anti-corporatism and environmental extremism to make their case. They will likely be opposed by the techno-libertarian and patient advocacy groups who presently support human cloning and embryonic stem cell research, and by the mainstream political establishment, at both the national and state levels, which sees nanotech as a way to boost the economy.

Just as there is no prominent figure in Washington arguing against nanotechnology, there is currently no prominent advocate of Eric Drexler's radical vision of nanotechnology. The closest thing to such an advocate may be Glenn Harlan Reynolds, a law professor from the University of Tennessee, whose Instapundit website and online columns are read by many Washingtonians. Reynolds frequently discusses nanotechnology, and when he does, he openly supports Drexler's ideas. (He also serves on the board of the Foresight Institute.) Reynolds is one of the few writers who understands both the workings of government and the basic theories of nanotechnology, which makes him useful to readers in the nation's capital.

If still unformed, however, there is reason to believe that public debate about nanotech is about to take off—with two new nanotech organizations founded in just the past year. The Center for Responsible Nanotechnology, run by a social activist and a nanosystems theorist, has been cranking out publications since January. "What we want," says Chris Phoenix, one of the Center's founders, "is to see molecular nanotechnology policy developed and implemented with a care appropriate to its powerful and probably transformative nature." And two Washingtonians—a futurist and an antitrust lawyer—are in the process of launching the Nanotechnology Policy Forum to improve the quality of public discourse about nanotech. They intend to host events every few months, and to stay scrupulously evenhanded: the advisory panel planned for the organization will include both friends and foes of nanotech—as well as present and former congressmen.

Congress also seems slightly more attuned to the need for debate about nanotechnology. Plans are afoot in both the House and the Senate to fund studies of the social, economic, and environmental implications of nanotechnology.

Also, legislation currently wending its way through Congress would establish "grand challenges" for nanotechnology: long-term objectives akin to President Kennedy's goal of putting a man on the Moon. While it isn't at all clear at this stage that nanotechnology can capture the imagination of the public like the Moon missions did, there is one obvious goal that would make an excellent "grand challenge"—a goal presently overlooked in all the millions of federal dollars going to nanotech: the assembler breakthrough. And just as the Apollo missions to the Moon were preceded by missions with incremental goals (achieved by the Mercury and Gemini programs), an ambitious nanotechnology project aspiring to make the world's first assembler could also set intermediate goals, like the creation of a basic nanoscale computer or a nanoscale robotic arm. But the National Nanotechnology Initiative is so focused on developing mainstream nanotech that Drexler's nanotechnology has found neither a great advocate nor a great critic.

Challenge Ahead

One Congressman, speaking on the House floor on May 7 of this year, said that nanotechnology may "create levels of intelligence that may be our protector, may be our competitor, or may simply regard us as pets. Or it may change our definition of what it is to be a human being." He said this with no indication of outrage or regret; he didn't rail against the prospect of a posthuman future; he expressed no aversion to life as a pet. Instead, he said that we need to talk about these issues, "to see how we can deal" with them, and to "get input from a wide range of society."

One proposal now under consideration is to create an advisory committee of "nonscientific and nontechnical" Americans to make recommendations about nanotechnology. This provision was inspired by the testimony of technology critic Langdon Winner before a House committee in April. "Congress should seek to create ways in which small panels of ordinary, disinterested citizens, selected in much the same way that we now choose juries in cases of law, [could] be assembled to examine important societal issues about nanotechnology," Winner said. The panels would listen to news and arguments, "deliberate on their findings, and write reports offering policy advice."

This sort of citizen panel would admittedly be an excellent gesture, a symbol of the fact that technological progress doesn't take place in a vacuum—that science and technology affect society as a whole and must remain subject to political oversight. But if the goal is really to inform the public and to get ordinary citizens to think about the implications of nanotechnology, a few citizen panels writing obscure reports won't have much effect. Nanotechnology education is most needed in newsrooms across the country and in the halls of the Capitol itself: We need reporters who know what they're talking about and who ask the right questions, and we need political leaders who can guide us through the confusing and potentially perilous times ahead.

This much seems clear: If molecular nanotechnology ever becomes a reality, we can expect massive social disruptions. As for the nature of these disruptions, we can, at best, only speculate.

First, we will hear complaints that the benefits of the new technology aren't being shared equitably; that the poor are being left out. But that problem will quickly fade away, as usually happens in our innovative market economy. (Think of the notorious "digital divide" of the 1990s, for instance; it's now all but gone.)

Second, we will reorganize our society and economy, shifting workers, eliminating jobs, and completely restructuring entire industries. We will hear questions about how our character will change: Will the abundance made possible by molecular manufacturing cause us to slip into hedonistic excess? Or might it have the reverse effect, making us *less* materialistic? Will life become so easy for us that it loses all meaning, or is it in our nature to keep seeking new challenges? (These questions may be as groundless as the "Leisure Question" that had social scientists wringing their hands a few decades ago, worrying about what we would do with all our free time in the age of automation and computers.)

Third, we will have to confront the "extinctionist" challenge and decide who we are. Nanotechnology raises many of the same ethical issues as biotechnology, and indeed the two techniques overlap. How much will we tinker with and revise our bodies? Will we choose a future as men or machines? Will we be able to use nanotechnology without drowning in nanotechnology, losing ourselves in nanotechnology, *becoming* nanotechnology?

And finally, we will be charged with rethinking our place in the universe. Our new powers of precision and perfection could lead us to a deeper appreciation for life—or they could make us lose all respect for the imperfect world we inhabit and the imperfect beings we have always been. The era of nanotechnology may be one of hubris and overreach, where we use our godlike powers to make the world anew. Is there room for wonder in a future where atoms march at our command?

Public debate about these matters will surely stay much lower to the ground—with arguments about where best to invest nanotech resources or about the quantifiable dangers of nanotechnology to the health and well-being of man and nature. Those who care about the deeper questions—about what nanotechnology means for human nature—must also master the details, both political and scientific. And they must offer not only lamentations for the disruptions and dehumanization that nanotechnology might cause, but a sensible vision of how nanotechnology might do some practical good—or even stir the very wonder that could be diminished by rearranging the smallest parts without seeing the whole.

10

A Door to the Future

Benjamin Franklin wanted a procedure for stopping and re starting metabolism, but none was then known. Do we live in a century far enough advanced to make biostasis available—to open a future of health to patients who would otherwise lack any choice but dissolution after they have expired?

We can stop metabolism in many ways, but biostasis, to be of use, must be reversible. This leads to a curious situation. Whether we can place patients in biostasis using *present* techniques depends entirely on whether *future* techniques will be able to reverse the process. The procedure has two parts, of which we must master only one.

If biostasis can keep a patient unchanged for years, then those future techniques will include sophisticated cell repair systems. We must therefore judge the success of present biostasis procedures in light of the ultimate abilities of future medicine. Before cell repair machines became a clear prospect, those abilities—and thus the requirements for successful biostasis—remained grossly uncertain. Now, the basic requirements seem fairly obvious.

Requirements for Biostasis

Molecular machines can build cells from scratch, as dividing cells demonstrate. They can also build organs and organ systems from scratch, as developing embryos demonstrate. Physicians will be able to use cell repair technology to direct the growth of new organs from a patient's own cells. This gives modern physicians great leeway in biostasis procedures: even if they were to damage or discard most of a patient's organs, they would still do no irreversible harm. Future colleagues with better tools will be able to repair or replace the organs involved. Most people would be glad to have a new heart, fresh kidneys, or younger skin.

But the brain is another matter. A physician who allows the destruction of a patient's brain allows the destruction of the patient as a person, whatever may happen to the rest of the body. The brain holds the patterns of memory, of personality, of self. Stroke patients lose only parts of their brains, yet suffer harm ranging from partial blindness to paralysis to loss of language, lowered intelligence, altered personality, and worse. The effects depend on the location of the damage. This suggests that total destruction of the brain causes total blindness, paralysis, speechlessness, and mindlessness, whether the body continues to breathe or not.

As Voltaire wrote, "To rise again—to be the same person that you were—you must have your memory perfectly fresh and present; for it is memory that makes your identity. If your memory be lost, how will you be the same man?" Anesthesia interrupts consciousness without disrupting the structure of the brain, and biostasis procedures must do likewise, for a longer time. This raises the question of the nature of the physical structures that underlie memory and personality.

Neurobiology, and informed common sense, agree on the basic nature of memory. As we form memories and develop as individuals, our brains change. These changes affect the brain's function, changing its pattern of activity: When we remember, our brains do some thing; when we act, think, or feel, our brains do something. Brains work by means of molecular machinery. Lasting changes in brain function involve lasting changes in this molecular machinery—unlike a computer's memory, the brain is not designed to be wiped clean and refilled at a moment's notice. Personality and long-term memory are durable.

Throughout the body, durable changes in function involve durable changes in molecular machinery. When muscles become stronger or swifter, their proteins change in number and distribution. When a liver adapts to cope with alcohol, its protein content also changes. When the immune system learns to recognize a new kind of influenza virus, protein content changes again. Since protein-based machines do the actual work of moving muscles, breaking down toxins, and recognizing viruses, this relationship is to be expected.

In the brain, proteins shape nerve cells, stud their surfaces, link one cell to the next, control the ionic currents of each neural impulse, produce the signal molecules that nerve cells use to communicate across synapses, and much, much more. When printers print words, they put down patterns of ink; when nerve cells change their behavior, they change their patterns of protein. Printing also dents the paper, and nerve cells change more than just their proteins, yet the ink on the paper and the proteins in the brain are enough to make these patterns clear. The changes involved are far from subtle. Researchers report that long-term changes in nerve cell behavior involve "striking morphological changes" in synapses: they change visibly in size and structure. It seems that long-term memory is not some terribly delicate pat tern, ready to evaporate from the brain at any excuse. Memory and personality are instead firmly embodied in the way that brain cells have grown together, in patterns formed through years of experience. Memory and personality are no more material than the characters in a novel; yet like them they are embodied in matter. Memory and personality do not waft away on the last breath as a patient expires. Indeed, many patients have recovered from so-called "clinical death," even without cell repair machines to help. The pat terns of mind are destroyed only when and if the attending physicians allow the patient's brain to undergo dissolution. This again allows physicians considerable leeway in biostasis procedures: typically, they need not stop metabolism until after vital functions have ceased.

It seems that preserving the cell structures and protein patterns of the brain will also preserve the structure of the mind and self. Biologists already know how to preserve tissue this well. Resuscitation technology must await cell repair machines, but biostasis technology seems well in hand.

Methods of Biostasis

The idea that we already have biostasis techniques may seem surprising, since powerful new abilities seldom spring up overnight. In fact, the techniques are old—only understanding of their reversibility is new. Biologists developed the two main approaches for other reasons.

For decades, biologists ha /e used electron microscopes to study the structure of cells and tissues. To prepare specimens, they use a chemical process called *fixation* to hold molecular structures in place. A popular method uses glutaraldehyde molecules, flexible chains of five carbon atoms with a reactive group of hydrogen and oxygen atoms at each end. Biologists fix tissue by pumping a glutaraldehyde solution through blood vessels, which allows glutaraldehyde molecules to diffuse into cells. A molecule tumbles around inside a cell until one end contacts a protein (or other reactive molecule) and bonds to it. The other end then waves free until it, too, contacts something reactive. This commonly shackles a protein molecule to a neighboring molecule.

These cross-links lock molecular structures and machines in place; other chemicals then can be added to do a more thorough or sturdy job. Electron microscopy shows that such fixation procedures preserve cells and the structures within them, including the cells and structures of the brain.

The first step of the hypothetical biostasis procedure involved simple molecular devices able to enter cells, block their molecular machinery, and tie structures together with stabilizing cross-links. Glutaraldehyde molecules fit this description quite well. The next step in this procedure involved other molecular devices able to displace water and pack themselves solidly around the molecules of a cell. This also corresponds to a known process.

Chemicals such as propylene glycol, ethylene glycol, and dimethyl sulfoxide can diffuse into cells, replacing much of their water yet doing little harm. They are known as "cryoprotectants," because they can protect cells from damage at low temperatures. If they replace enough of a cell water, then cooling doesn't cause freezing, it just causes the protectant solution to become more and more viscous, going from a liquid that resembles thin syrup in its consistency to one that resembles hot tar, to one that resembles cold tar, to one as resistant to flow as a glass. In fact, according to the scientific definition of the term, the protectant solution then qualifies as a glass; the process of solidification without freezing is called vitrification. Mouse embryos vitrified and stored in liquid nitrogen have grown into healthy mice.

The vitrification process packs the glassy protectant solidly around the molecules of each cell; vitrification thus fits the description I gave of the second stage of biostasis.

Fixation and vitrification together seem adequate to ensure long-term biostasis. To reverse this form of biostasis, cell repair machines will be programmed to remove the glassy protectant and the glutaraldehyde cross-links and then repair and replace molecules, thus restoring cells, tissues, and organs to working order.

Fixation with vitrification is not the first procedure proposed for biostasis. In 1962 Robert Ettinger, a professor of physics at Highland Park College in Michigan, published a book suggesting that future advances in cryobiology might lead to techniques for the easily reversible freezing of human patients. He further suggested that physicians using future technology might be able to repair and revive patients frozen with present techniques shortly after cessation of vital signs. He pointed out that liquid nitrogen temperatures will preserve patients for centuries, if need be, with little change. Perhaps, he suggested, medical science will one day have "fabulous machines" able to restore frozen tissue a molecule at a time. His book gave rise to the cryonics movement.

Cryonicists have focused on freezing because many human cells revive *spontaneously* after careful freezing and thawing. It is a common myth that freezing bursts cells; in fact, freezing damage is more subtle than this—so subtle that it often does no lasting harm. Frozen sperm regularly produces healthy babies. Some human beings now alive have survived being frozen solid at liquid nitrogen temperatures—when they were early embryos. Cryobiologists are actively researching ways to freeze and thaw viable organs to allow surgeons to store them for later implantation.

The prospect of future cell repair technologies has been a consistent theme among cryonicists. Still, they have tended to focus on procedures that preserve cell function, for natural reasons. Cryobiologists have kept viable human cells frozen for years. Researchers have improved their results by experimenting with mixes of cryoprotective chemicals and carefully controlled cooling and warming rates. The complexities of cryobiology offer rich possibilities for further experimentation. This combination of tangible, tantalizing success and promising targets for further research has made the quest for an easily reversible freezing process a vivid and attractive goal for cryonicists. A success at freezing and reviving an adult mammal would be immediately visible and persuasive.

What is more, even *partial* preservation of tissue function suggests *excellent* preservation of tissue structure. Cells that can revive (or almost revive) even without special help will need little repair.

The cryonics community's cautious, conservative emphasis on pre serving tissue function has invited public confusion, though. Experimenters have frozen whole adult mammals and thawed them without waiting for the aid of cell repair machines. The results have been superficially discouraging: the animals fail to revive. To a public and a medical community that has known nothing about the prospects for cell repair, this has made frozen biostasis seem pointless.

And, after Ettinger's proposal, a few cryobiologists chose to make unsupported pronouncements about the future of medical technology. As Robert Prehoda stated in a 1967 book: "Almost all reduced metabolism experts ... believe that cellular damage caused by cur rent freezing techniques could never be corrected." Of course, these were the wrong experts to ask. The question called for experts on molecular technology and cell repair machines. These cryobiologists should have said only that correcting freezing damage would apparently require molecular-level repairs, and that they, personally, had not studied the matter. Instead, they casually misled the public on a matter of vital medical importance. Their statements discouraged the use of a workable biostasis technique.

Cells are mostly water. At low enough temperatures, water molecules join to form a weak but solid framework of cross-links. Since this preserves neural structures and thus the patterns of mind and memory, Robert Ettinger has apparently identified a workable approach to biostasis. As molecular technology advances and people grow familiar with its consequences, the reversibility of biostasis (whether based on freezing, fixation and vitrification, or other methods) will grow ever more obvious to ever more people.

Reversing Biostasis

Imagine that a patient has expired because of a heart attack. Physicians attempt resuscitation but fail, and give up on restoring vital functions. At this point, though, the patient's body and brain are just *barely* nonfunctional—most cells and tissues, in fact, are still alive and metabolizing. Having made arrangements beforehand, the patient is soon placed in biostasis to prevent irreversible dissolution and await a better day.

Years pass. The patient changes little, but technology advances greatly. Biochemists learn to design proteins. Engineers use protein machines to build assemblers, then use assemblers to build a broad-based nanotechnology. With new instruments, biological knowledge explodes. Biomedical engineers use new knowledge, automated engineering, and assemblers to develop cell repair machines of growing sophistication. They learn to stop and reverse aging. Physicians use cell repair technology to resuscitate patients in biostasis—first those placed in biostasis by the most advanced techniques, then those placed in biostasis using earlier and cruder techniques. Finally, after the successful resuscitation of animals placed in biostasis using the old techniques of the 1980s, physicians turn to our heart-attack patient.

In the first stage of preparation, the patient lies in a tank of liquid nitrogen surrounded by equipment. Glassy protectant still locks each cell's molecular machinery in a firm embrace. This protectant must be removed, but simple warming might allow some cell structures to move about prematurely.

Surgical devices designed for use at low temperatures reach through the liquid nitrogen to the patient's chest. There they remove solid plugs of tissue to open access to major arteries and veins. An army of nano machine equipped for removing protectant moves through these openings, clearing first the major blood vessels and then the capillaries. This opens paths throughout the normally active tissues of the patient's body. The larger surgical machines then attach tubes to the chest and pump fluid through the circulatory system. The fluid washes out the initial protectant-removal machines (later, it supplies materials to repair machines and carries away waste heat).

Now the machines pump in a milky fluid containing trillions of devices that enter cells and remove the glassy protectant, molecule by molecule. They replace it with a temporary molecular scaffolding that leaves ample room for repair machines to work. As these protectant machines uncover biomolecules, including the structural and mechanical components of the cells, they bind them to the 5 with temporary cross-links. (If the patient had also been treated with a cross-linking fixative, these cross-links would now be removed and replaced with the temporary links.) When molecules must be moved aside, the machines label them for proper replacement. Like other advanced cell repair machines, these devices work under the direction of on-site nanocomputers.

When they finish, the low-temperature machines withdraw. Through a series of gradual changes in composition and temperature, a water solution replaces the earlier cryogenic fluid and the patient warms to above the freezing point. Cell repair machines are pumped through the blood vessels and enter the cells. Repairs commence.

Small devices examine molecules and report their structures and positions to a larger computer within the cell. The computer identities the molecules, directs any needed molecular repairs, and identities cell structures from molecular patterns. Where damage has displaced structures in a cell, the computer directs the repair devices to restore the molecules to their proper arrangement, using temporary cross-links as needed. Meanwhile, the patient's arteries are cleared and the heart muscle, damaged years earlier, is repaired.

Finally, the molecular machinery of the cells has been restored to working order, and coarser repairs have corrected damaged pattern of cells to restore tissues and organs to a healthy condition. The scaffolding is then removed from the cells, together with most of the temporary cross-links and much of the repair machinery. Most of each cell's active molecules remain blocked, though, to prevent pre mature, unbalanced activity.

Outside the body, the repair system has grown fresh blood from the patient's own cells. It now transfuses this blood to refill the circulatory system, and acts as a temporary artificial heart. The remaining devices in each cell now adjust the concentration of salts, sugars, ATP, and other small molecules, largely by selectively unblocking each cell's own nanomachinery. With further unblocking, metabolism resumes step by step; the heart muscle is finally unblocked on the verge of contraction. Heartbeat resumes, and the patient emerges into a state of anesthesia. While the attending physicians check that all is going well, the repair system closes the opening in the chest, joining tissue to tissue without a stitch or a scar. The remaining devices in the cells disassemble one another into harmless waste or nutrient molecules. As the patient moves into ordinary sleep, certain visitors enter the room, as long planned.

At last, the sleeper wakes refreshed to the light of a new day—and to the sight of old friends.

Mind, Body, and Soul

Before considering resuscitation, however, some may ask what be comes of the soul of a person in biostasis. Some people would answer that the soul and the mind are aspects of the same thing, of a pattern embodied in the substance of the brain, active during active life and quiescent in biostasis. Assume, though, that the pattern of mind, memory, and personality leaves the body at death, carried by some subtle substance. The possibilities then seem fairly clear. Death in this case has a meaning other than irreversible damage to the brain, being defined instead by the irreversible departure of the soul. This would make biostasis a pointless but harmless gesture—after all, religious leaders have expressed no concern that mere preservation of the body can somehow imprison a soul. Resuscitation would, in this view, presumably require the cooperation of the soul to succeed. The act of placing patients in biostasis has in fact been accompanied by both Catholic and Jewish ceremonies.

With or without biostasis, cell repair cannot bring immortality. Physical death, however greatly postponed, will remain inevitable for reasons rooted in the nature of the universe. Biostasis followed by cell repair thus seems to raise no fundamental theological issues. It resembles deep anesthesia followed by life-saving surgery: both procedures interrupt consciousness to prolong life. To speak of "immortality" when the prospect is only long life would be to ignore the facts or to misuse words.

Reactions and Arguments

The prospect of biostasis seems tailor-made to cause future shock. Most people find today's accelerating change shocking enough when it arrives a bit at a time. But the biostasis option is a present-day consequence of a whole series of future breakthroughs. This prospect naturally upsets the difficult psychological adjustments that people make in dealing with physical decline.

Thus far, I have built the case for cell repair and biostasis on a discussion of the commonplace facts of biology and chemistry. But what do professional biologists think about the basic issues? In particular, do they believe (1) that repair machines will be able to correct the kind of cross-linking damage produced by fixation, and (2) that memory is indeed embodied in a preservable form?

After a discussion of molecular machines and their capabilities—a discussion not touching on medical implications—Dr. Gene Brown, professor of biochemistry and chairman of the department of biology at MIT, gave permission to be quoted as stating that: "Given sufficient time and effort to develop artificial molecular machines and to conduct detailed studies of the molecular biology of the cell, very broad abilities should emerge. Among these could be the ability to separate the proteins (or other biomolecules) in cross-linked structures, and to identify, repair, and replace them." This statement ad dresses a significant part of the cell repair problem. It was consistently endorsed by a sample of biochemists and molecular biologists at MIT and Harvard after similar discussion.

After a discussion of the brain and the physical nature of memory and personality—again discussion not touching on medical implications—Dr. Walle Nauta (Institute Professor of Neuroanatomy at MIT) gave permission to be quoted as stating that: "Based on our present knowledge of the molecular biology of neurons, I think most would agree that the changes produced during the consolidation of long-term memory are reflected in corresponding changes in the number and distribution of different protein molecules in the neurons of the brain." Like Dr. Brown's statement, this addresses a key point regarding the workability of biostasis. It, too, was consistently endorsed by other experts when discussed in a context that insulated the experts from any emotional bias that might result from the medical implications of the statement. Further, since these points relate directly to their specialties, Dr. Brown and Dr. Nauta were appropriate experts to ask.

It seems that the human urge to live will incline many millions of people toward using biostasis (as a last resort) if they consider it workable. As molecular technology advances, understanding of cell repair will spread through the popular culture. Expert opinion will increasingly support the idea. Biostasis will grow more common, and its costs will fall. It seems likely that many people will eventually consider biostasis to be the norm, to be a standard lifesaving treatment for patients who have expired.

But until cell repair machines are demonstrated, the all too human tendency to ignore what we have not seen will slow the acceptance of biostasis. Millions will no doubt pass from expiration to irreversible dissolution because of habit and tradition, supported by weak arguments. The importance of clear foresight in this matter makes it important to consider possible arguments before leaving the topic of life extension and moving on to other matters. Why, then, might biostasis not seem a natural, obvious idea?

It may seem strange to save a person from dissolution in the expectation of restoring health, since the repair technology doesn't exist yet. But is this so much stranger than saving money to put a child through college? After all, the college student doesn't exist yet, either. Saving money makes sense because the child will mature; saving a person makes sense because molecular technology will mature.

We expect a child to mature because we have seen many children mature; we can expect this technology to mature because we have seen many technologies mature. True, some children suffer from congenital shortcomings, as do some technologies, but experts often can estimate the potential of children or technologies while they remain young.

Microelectronic technology started with a few spots and wires on a chip of silicon, but grew into computers on chips. Physicists such as Richard Feynman saw, in part, how far it would lead.

Nuclear technology started with a few atoms splitting in the laboratory under neutron bombardment, but grew into billion-watt reactors and nuclear bombs. Leo Szilard saw, in part, how far it would lead.

Liquid rocket technology started with crude rockets launched from a Massachusetts field, but grew into Moon ships and space shuttles.

Robert Goddard saw, in part, how far it would lead.

Molecular engineering has started with ordinary chemistry and molecular machines borrowed from cells, but it, too, will grow mighty. It, too, has discernible consequences.

We tend to expect dramatic results only from dramatic causes, but the world often fails to cooperate. Nature delivers both triumph and disaster in brown paper wrappers.

Dull fact: Certain electric switches can turn one another on and off. These switches can be made very small, and frugal of electricity.

The dramatic consequence: When properly connected, these switches form computers, the engines of the information revolution.

Dull fact: Ether is not too poisonous yet temporarily interferes with the activity of the brain.

The dramatic consequence: An end to the agony of surgery on conscious patients, opening a new era in medicine.

Dull fact: Molds and bacteria compete for food, so some molds have evolved to secrete poisons that kill bacteria.

The dramatic consequence: Penicillin, the conquest of many bacterial diseases, and the saving of millions of lives.

Dull fact: Molecular machines can be used to handle molecules and build mechanical switches of molecular size.

The dramatic consequence: Computer cell repair machines, bringing cures for virtually all diseases.

Dull fact: Memory and personality are embodied in preservable brain structures.

The dramatic consequence: Present techniques can prevent dissolution, letting the present generation take advantage of tomorrow's cell repair machines.

In fact, molecular machines aren't even so dull. Since tissues are made of atoms, one should expect a technology able to handle and rearrange atoms to have dramatic medical consequences.

In an article entitled "The Idea of Progress" in *Astronautics* and *Aeronautics*, aerospace engineer Robert T, Jones wrote: "In 1950, the year I was born, my father was a prosecuting attorney. He traveled all the dirt roads in Macon County in a buggy behind a single horse. Last year I flew nonstop from London to San Francisco over the polar regions, pulled through the air by engines of 50,000 horsepower." In his father's day, such aircraft lay at the fringe of science fiction, too incredible to consider.

In an article entitled "Basic Medical Research: A Long-Term In vestment" in MIT's Technology Review, Dr. Lewis Thomas wrote: "Forty years ago, just before the profession underwent transformation from an art to science and technology, it was taken for granted that the medicine we were being taught was precisely the medicine that would be with us for most of our lives. If anyone had tried to tell us that the power to control bacterial infections was just around the corner, that open-heart surgery or kidney transplants would be possible within a couple of decades, that some kinds of cancer could be cured by chemotherapy, and that we would soon be within reach of a comprehensive, biochemical explanation for genetics and genetically determined diseases, we would have reacted in blank disbelief. We had no reason to believe that medicine would ever change. What this recollection suggests is that we should keep our minds wide open in the future."

News of a way to avoid the fatality of most fatal diseases may indeed sound too good to be true—as it should, since it is but a small part of a more balanced story. In fact, the dangers of molecular technology roughly balance its promise. In Part Three I will outline reasons for considering nanotechnology more dangerous than nuclear weapons.

Fundamentally, though, nature cares nothing for our sense of good and bad and nothing for our sense of balance. In particular, nature does not hate human beings enough to stack the deck against us. Ancient horrors have vanished before.

Years ago, surgeons strove to amputate legs fast. Robert Liston of Edinburgh, Scotland, once sawed through a patient's thigh in a record thirty-three seconds, removing three of his assistant's fingers in the process. Surgeons worked fast to shorten their patients' agony, because their patients remained conscious.

If terminal illness without biostasis is a nightmare today, consider surgery without anesthesia in the days of our ancestors: the knife slicing through flesh, the blood flowing, the saw grating on the bone of a conscious patient. ... Yet in October of 1846, W. T. G. Morton and J. C. Warren removed a tumor from a patient under ether anesthesia; Arthur Slater states that their success "was rightly hailed as the great discovery of the age." With simple techniques based on a known chemical, the waking nightmare of knife and saw at long last was ended.

With agony ended, surgery increased, and with it surgical infection and the horror of routine death from flesh rotting in the body. Yet in

1867 Joseph Lister published the results of his experiments with phenol, establishing the principles of antiseptic surgery. With simple techniques based on a known chemical, the nightmare of rotting alive shrank dramatically.

Then came sulfa drugs and penicillin, which ended many deadly diseases in a single blow ... the list goes on.

Dramatic medical breakthroughs have come before, sometimes from new uses of known chemicals, as in anesthesia and antiseptic surgery. Though these advances may have seemed too good to be true, they were true nonetheless. Saving lives by using known chemicals and procedures to produce biostasis can likewise be true.

Robert Ettinger proposed a biostasis technique in 1962. He states that Professor Jean Rostand had proposed the same approach years earlier, and had predicted its eventual use in medicine. Why did biostasis by freezing fail to become popular? In part because of its initial expense, in part because of human inertia, and in part because means for repairing cells remained obscure. Yet the ingrained conservatism of the medical profession has also played a role. Consider again the history of anesthesia.

In 1846, Morton and Warren amazed the world with the "discovery of the age," ether anesthesia. Yet two years earlier, Horace Wells had used nitrous oxide anesthesia, and two years before that, Craw ford W. Long had performed an operation using ether. In 1824, Henry Hickman had successfully anesthetized animals using ordinary carbon dioxide; he later spent years urging surgeons in England and France to test nitrous oxide as an anesthetic. In 1799, *a full forty-seven years before the great "discovery,"* and years before Liston's assistant lost his fingers, Sir Humphry Davy wrote: "As nitrous oxide in its extensive operation appears capable of destroying physical pain, it may possibly be used during surgical operations."

Yet as late as 1839 the conquest of pain still seemed an impossible dream to many physicians. Dr. Alfred Velpeau stated: "The abolishment of pain in surgery is a chimera. It is absurd to go on seeking it today. 'Knife' and 'pain' are two words in surgery that must forever be associated in the consciousness of the patient. To this compulsory combination we shall have to adjust ourselves."

Many feared the pain of surgery more than death itself. Perhaps the time has come to awaken from the final medical nightmare.

It is true that no experiment can now demonstrate the resuscitation of a patient in biostasis. But a demand for such a demonstration would carry the hidden assumption that modem medicine has neared the final limits of the possible, that it will never be humbled by the achievements of the future. Such a demand might sound cautious and reasonable, but in fact it would smack of overwhelming arrogance.

Unfortunately, a demonstration is exactly what physicians have been trained to request, and for good reason: they wish to avoid useless procedures that may do harm. Perhaps it will suffice that neglect of biostasis leads to obvious and irreversible harm.

Time, Cost, and Human Action

Whether people choose to use biostasis will depend on whether they see it as worth the gamble. This gamble involves the value of life (which is a personal matter), the cost of biostasis (which seems reasonable by the standards of modern medicine), the odds that the technology will work (which seem excellent), and the odds that humanity will survive, develop the technology, and revive people. This final point accounts for most of the overall uncertainty.

Assume that human beings and free societies will indeed survive. (No one can calculate the odds of this, but to assume failure would discourage the very efforts that will promote success.) If so, then technology will continue to advance. Developing assemblers will take years. Studying cells and learning to repair the tissues of patients in biostasis will take still longer. At a guess, developing repair systems and adapting them to resuscitation will take three to ten decades, though advances in automated engineering may speed the process.

The time required seems unimportant, however. Most resuscitated patients will care more about the conditions of life—including the presence of their friends and family—than they will care about the date on the calendar. With abundant resources, the physical conditions of life could be very good indeed. The presence of companions is another matter.

In a recently published survey, over half of those responding said that they would like to live for at least five hundred years, if given a free choice. Informal surveys show that most people would prefer biostasis to dissolution, if they could regain good health and explore a new future with old companions. A few people say that they "want to go when their time comes," but they generally agree that, so long as they can choose further life, their time has not yet come. It seems that many people today share Benjamin Franklin's desire, but in a century able to satisfy it. If biostasis catches on fast enough (or if other life-extension technologies advance fast enough), then a resuscitated patient will awake not to a world of strangers, but to the smiles of familiar faces.

But will people in biostasis be resuscitated? Techniques for placing patients in biostasis are already known, and the costs could become low, at least compared to the costs of major surgery or prolonged hospital care. Resuscitation technology, though, will be complex and expensive to develop. Will people in the future bother?

It seems likely that they will. They may not develop nanotechnology with medicine in mind—but if not, then they will surely develop it to build better computers. They may not develop cell repair machines with resuscitation in mind, but they will surely do so to heal themselves. They may not program repair machines for resuscitation as an act of impersonal charity, but they will have time, wealth, and automated engineering systems, and some of them will have loved ones waiting in biostasis. Resuscitation techniques seem sure to be developed.

With replicators and space resources, a time will come when people have wealth and living space over a thousand fold greater than we have today. Resuscitation itself will require little energy and material even by today's standards. Thus, people contemplating resuscitation will find little conflict between their self-interest and their humanitarian concerns. Common human motives seem enough to ensure that the active population of the future will awaken those in biostasis.

The first generation that will regain youth without being forced to resort to biostasis may well be with us today. The prospect of biostasis simply gives more people more reason to expect long life—it offers an opportunity for the old and a form of insurance for the young. As advances in biotechnology lead toward protein design, assemblers, and cell repair, and as the implications sink in, the expectation of long life will spread. By broadening the path to long life, the biostasis option will encourage a more lively interest in the future. And this will spur efforts to prepare for the dangers ahead.

11

MANIPULATING TECHNIQUES

The manipulation of gene is the called the genetic engineering. Genetics is also important in manipulating biological systems for scientific or economic reasons, an endeavor that has made possible the creation of organisms having new phenotypes or genotypes. The fundamental techniques for accomplishing this have been mutagenesis or recombination followed by selection for desired characteristics. When using such techniques, geneticists have been forced to work with the random nature of mutagenic and recombination events, which required selective procedures, often quite complex, to find an organism with the required genotype among the many types of organisms produced. Since the 1970s techniques have been developed by which the genotype of an organism can instead be modified in a *directed* and *predetermined* way. This is alternately called *recombinant DNA technology*, or *genetic engineering*.

Genetic engineering involves isolation of DNA fragments and recombination outside of a cell. Selection of a desired genotype is still necessary but the probability of success is usually many orders of magnitude greater than that with traditional procedures. The basic technique is quite simple: two DNA molecules are isolated and cut into fragments by one or more specialized enzymes and then the fragments are joined together *in a desired combination* and restored to a cell for replication and reproduction.

Current interest in genetic engineering centers on its many practical applications, a few examples of which are the following:

1. Isolation of a particular gene, part of a gene, or region of a genome.
2. Production of particular RNA and protein molecules in quantities formerly thought to be unobtainable.

3. Improvement in the production of biochemicals (such as enzymes and drugs) and commercially important organic chemicals.
4. Production of varieties of plants having particular desirable characteristics (for example, requiring less fertilizer or resistance to disease).
5. Correction of genetic defects in higher organisms, and
6. Creation of organisms with economically important features (for example, plants capable of maturing faster or having greater yield). Some of these examples will be considered later in the chapter.

Separation of Particular DNA Fragments

In genetic engineering the immediate goal of an experiment is usually to insert a *particular* fragment of chromosomal DNA into a plasmid or a viral DNA molecule. This is accomplished by techniques for breaking DNA molecules at specific sites and for isolating particular DNA fragments. DNA fragments are obtained by treatment of DNA samples with a specific class of nuclease. Many nucleases have been isolated from a variety of organisms, and most produce breaks at random sites within a DNA sequence. However, the class of nucleases called *restriction endonucleases*, or, more simply, *restriction enzymes*, consists of sequence-specific enzymes. Most restriction enzymes recognize only one short base sequence in a DNA molecule and make two single-strand breaks, one in each strand, generating 3′-OH and 5′-P groups at each position. Several hundred of these enzymes have been isolated from hundreds of species of microorganisms.

The sequences recognized by restriction enzymes are often *palindromes*—that is, the sequence has symmetry of the form:

A B C	C′ B′ A′		A B	X	B′ A′		A B	B′ A′
A′ B′ C′	C B A	or	A′ B′	X′	B A	or	A′ B′	B A

in which the capital letter represent bases, a′ indicates a complementary base, X is any base, and the vertical line is the axis of symmetry. Most of these sequence have 4-6 bases.

One of the most exciting events in the study of restriction enzymes was the observation by electron microscopy that fragments produced by many restriction enzymes spontaneously circularize. These circles could be relinearized by heating, but if after circularization they were also treated with *E. coli* DNA ligase, which joins 3′-OH and 5′-P groups, circularization became permanent. This observation was the first evidence for three important features of restriction enzymes:

Table 11.1. Some restriction endonucleases, their sources, and their cleavage sites.

Name of enzyme	*Microorganism*	*Target sequence and cleavage sites*
Generates cohesive ends		
EcoRI	*E.coli*	G↓A A | T T C C T T | A A↑G
BamHI	*Bacillus amyloliquefaciens* H	G↓G A | T C C C C T | A G↑G
HaeII	*Haemophilus aegyptius*	Pu G C | G C↓Py Py↑C G | C G Pu
HindIII	*Haemophilus influenza*	A↓A G | C T T T T C | G A↑A
PstI	*Providencia stuartii*	C T G | C A↓G G A C | G T C
TaqI	*Thermus aquaticus*	T↓C | G A A G | C↑T
Generates blunt ends		
BalI	*Brevibacterium albium*	T G G ↓ C C A A C C ↑ G G T
SmaI	*Serratia marcescens*	C C C ↓ G G G G G G ↑ C C C

Note : The vertical dashed line indicates the axis of symmetry in each sequence. Arrows indicate the sites of cutting. The enzyme TaqI yields cohesive ends consisting of two nucleotides, whereas the cohesive ends produced by the other enzymes contain four nucleotides. Pu and Py refer to any purine and pyrimidine, respectively.

1. Restriction enzymes make breaks in palindromic sequences.
2. The breaks are usually not directly opposite one another.
3. The enzymes generate DNA fragments with complementary ends.

Examination of a very large number of restriction enzymes showed that the breaks are usually in one of two distinct arrangements: (1)

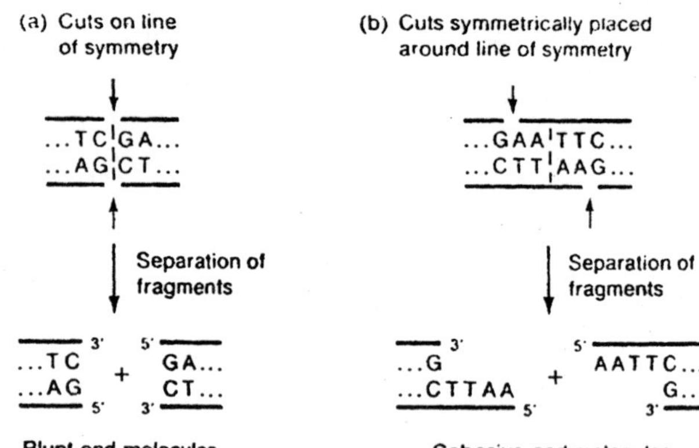

Fig. 11.1. Two types of cuts made by restriction enzymes. The arrow indicate the cleavage sites. The dashed line is the center of symmetry of the sequence.

staggered, but symmetric around the line of symmetry (forming *cohesive ends*) or (2) both at the center of symmetry (forming *blunt ends*). Two types of enzymes produce cohesive ends—those yielding a single-stranded extension with a 5′-P terminus and those yielding a 3′-OH extension. Table 13.1. lists the sequences and cleavage sites for several restriction enzymes, some of which generate cohesive sites and others of which yield blunt ends.

Most restriction enzymes recognize one base sequence without regard to the source of the DNA. Thus,

> Fragments obtained from a DNA molecule from one organism have the same cohesive ends as the fragments produced by the same enzyme acting on DNA molecules from another organism.

Since most restriction enzymes recognize a unique sequence, *the number of cuts made in the DNA from an organism by a particular enzyme is limited.* A typical bacterial DNA molecule, which contains roughly 3×10^6 base pairs, is cut into several hundred to several thousand fragments, and nuclear DNA of mammals is cut into more than a million fragments. These numbers are large but still small compared to the number of sugar-phosphate bonds in an organism. Of special interest are the smaller DNA molecules, such as viral or plasmid DNA, which may have only 1-10 sites of cutting (or even none) for particular enzymes.

Plasmids having a single site for a particular enzyme are especially valuable. Because of the sequence specificity, *a particular restriction enzyme generates a unique set of fragments for a particular DNA molecule.* Another enzyme will generate a different set of fragments from the same DNA molecule. Figure (given below) shows the sites of cutting of *E. coli* phage λ DNA by the enzymes EcoRI and BamHI. A map showing the unique sites of cutting of the DNA of a particular organism by a single enzyme is called a *restriction map*.

The family of fragments generated by a single enzyme can be detected easily by gel electrophoresis of enzyme-treated DNA, and particular DNA fragments can be isolated by cutting out the portion of the gel containing the fragment and removing the DNA from the gel. Several techniques enable one to locate particular genes on fragments of a restriction map. One of the most generally applicable procedures is *Southern blotting*. In this procedure a gel in which DNA molecules have been separated by electrophoresis is treated with alkali to render the DNA single-stranded (denature the DNA) and then the DNA is transferred to a sheet of nitrocellulose in a way that the relative positions of the DNA bands are maintained. The nitrocellulose, to which the single-stranded DNA tightly binds, is then exposed to radioactive RNA or DNA in a way that leads to renaturation.

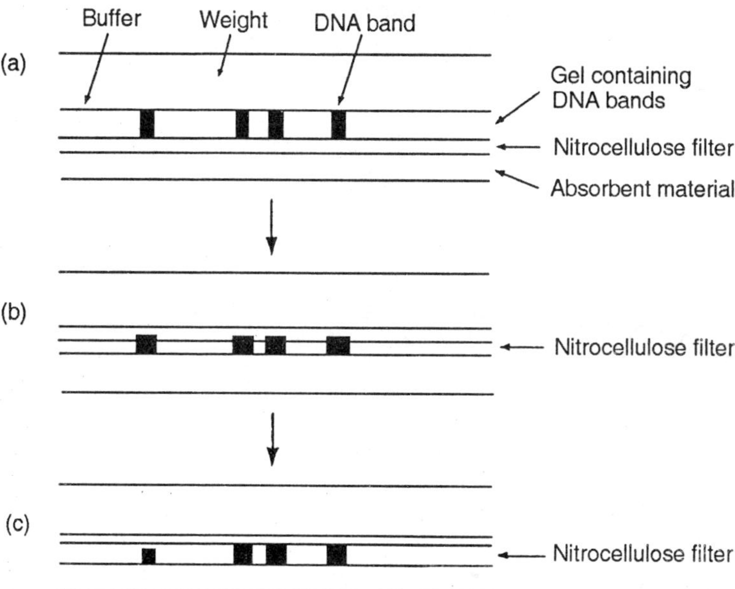

Fig. 11.2. Southern blotting. (a) A weight is placed on a stack consisting of the gel, a nitrocellulose filter, and absorbent material. (b) At a later time the weight has forced the buffer, which carries the DNA, into the nitrocellulose. (c) The lowest layer has absorbed the buffer but the DNA, which remains bound to the nitrocellulose.

Radioactivity becomes stably bound (resistant to removal by washing) to the DNA only at positions at which base sequences complementary to the radioactive molecules are present. The radioactivity is located by placing the paper in contact with x-ray film; after development of the film, blackened regions indicate positions of radioactivity. If a radioactive mRNA species transcribed from a particular gene is used (for example, mRNA isolated from a specialized cell that predominantly makes one type of mRNA), it will hybridize only with the restriction fragment containing that gene.

Restriction Endonucleases

Restriction enzymes are nucleases and as they cut at an internal position of DNA strand (and not at end) they are known as endonucleases. Restriction enzymes fall into following categories:

Type I : These are most complex, bi-functional (i.e. same enzyme possesses both restriction and modification activity) and are made up of 3 subunits. Recognition site is bipartite and asymmetrical. Cleavage site is nonspecific and is atleast 1000 bp from recognition site.

Type II : These are simplest, separate endonuclease and methylase. Recognition site is short sequence (4-6 bp) often palindromic

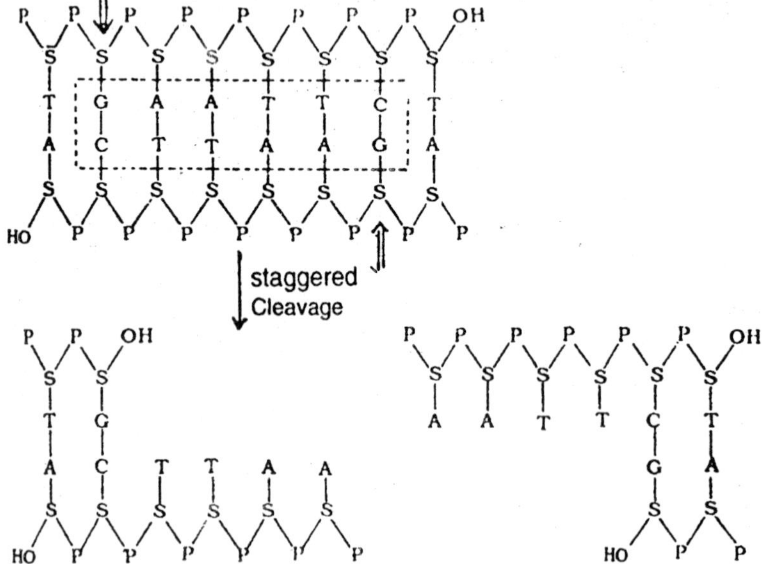

Fig. 11.3. Symmetrical staggered cleavage by Eco RI arrows indicate sites of cleavage.

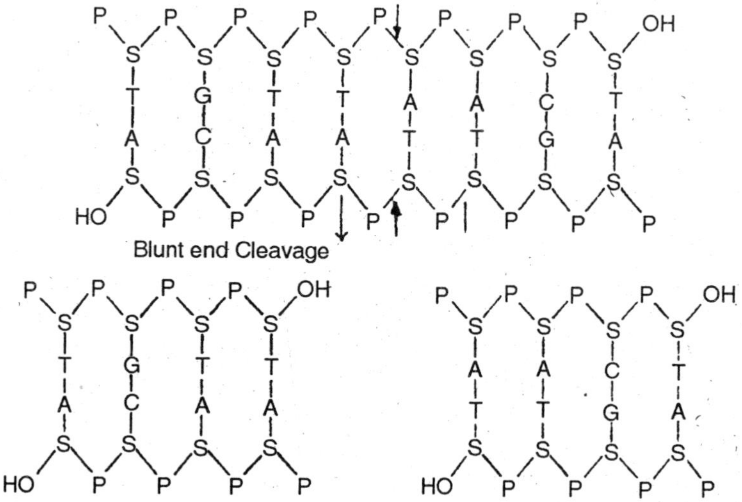

Blunt end Cleavage

Fig. 11.4. Blunt end cleavage by hind III.

(with symmetry). Cleavage site is same as or close to recognition site.

Type III : These are moderate complex, bi-functional. have two subunits. Recognition site is asymmetrical sequence of 5-7 bp. Cleavage site is 24-26 bp downstream from recognition site.

Type II restriction endonucleases are most important tools in gene manipulation techniques. Most of the restriction enzymes cleave DNA at unmethylated target site. Discovery of restriction enzymes (restriction endonucleases) has been useful in most of the modern genetic manipulations. Restriction endonucleases are indispensable tools for recombinant DNA research. They recognize specific oligonucleotide sequence and make double stranded cleavage to generate unique fragment of DNA molecule. Restriction endonucleases are used for dissecting, analysing and re-configuring genetic information at molecular level. Restriction endonucleases are used in: (1) cleavage mapping, (2) preparation of DNA probes, (3) gene cloning etc.

Nomenclature

There are over 2400 type II restriction endonucleases known so far (since 1968) and they recognize more than 188 different restriction sites. These enzymes have been well characterized.

The nomenclature of restriction enzyme is based on various conventions: (1) The generic and specific name of organism in which the enzyme is found are used to provide first part of the designation. Next number in designation indicates serial number or enzyme reported from same organism. Three letter abbrevation is in italic. (2) Strain or type identification is written as a subscript e.g. Eco_k. If restriction and modification system is genetically specified virus or plasmid, the abbreviated species name of host is given and extrachromosomal element as subscript. (3) When a particular host strain has several different restriction and modification system, these are identified by roman numericals I, II, III etc. following 3 letter abbreviation of host strain. (4) The history and incompleteness of the discovery are reflected in nomenclature of restriction endonucleases.

Unidentified bacteria found with these systems receive the genus-species designate *Uba*. As on 1993 hundreds of enzymes were temporarily named as *Uba* (number). In 1973, nomenclature of restriction enzyme was developed and was based on proposals of Smith and Nathan.

Recognition Sites

Restriction enzymes usually recognize a specific DNA sequence of 4, 5 or 6 nucleotides in length and cleave the DNA within this restriction site. There are 4 bases in DNA, randomly distributed. The expected frequency of any particular sequence can be calculated as 4^n where n is the length of recognition sequence. Thus, tetranucleotide sites will occur every 256 basepair, pentanucleotide sites will occur every 1025 basepairs and hexanucleotide sites will occur every 4096 basepairs. Restriction enzymes either cut (1) Straight across the DNA to give blunt ends, or (2) Straight single strand cuts producing short, single stranded projections at each end of the cleaved DNA to produce cohesive or sticky ends.

Table 11.2. Classification of restriction endonucleases based on recognition site.

(I) Tetranucleotides recognizing	—	*Alu I, Hae III, Hpa II, Mbo I, Taq I, Msp I* etc.
(II) Pentanucleotides recognizing	—	*Ava II, Dde I, Eco R II, Hinf I* etc.
(III) Hexanucleotides recognizing	—	*Kpn I, Xma I, Pst I, Hpal, Bam HI, Hind III* etc.
(IV) Heptanucleotides recognizing	—	*Me II* etc.

One unit of enzyme (Restriction endonuclease) is defined as the amount of enzyme required to produce a complete digest of 1 μg of substrate DNA in a reaction volume of 0.05 ml in 1 hour under optimal conditions of salt, pH and temperature. All digestions are performed at 37°C and substrate used is lambda DNA. Optimum temperature for most of the restriction enzymes is 37°C. However, in some cases like *Tha I, Taq I, Bcl I, Bst NI* optimum temperature is 60-65°C. Recognition sequences of *Eco RI* are *palindromic*. Sometimes within a same strand sequence may be palindromic.

Table 11.3. Restriction endonucleases.

	Enzyme	*Source*	*5´ Recognition site 3´*
1.	*Kpn II*	*Klebsiella pneumoniae*	GGTAC↓C
2.	*Bam HI*	*Bacillus amyloliquifaciens*	G↓GATCC
3.	*Sma I*	*Serratia marcesecens*	CCC↓GGG
4.	*Stu I*	*Streptomyces tubercidicus*	AGG↓CCT
5.	*AVA II*	*Anabenia variabilis*	G↓G(A/T) CC
6.	*Pva I*	*Proteus vulgaris*	CGAT↓CG
7.	*Hind III*	*Hemophilus influenzae Rd*	A↓AGCTT
8.	*Rsa I*	*Rhodopseudomonas sphaeroids*	GT↓AC
9.	*Hin C II*	*Hemophilus influenzae C1161*	GTPyPvAC
10.	*ECORI*	*E.coli RY 13*	G↓AATTC
11.	*Xma I*	*Xanthomonas malvacearum*	C↓CCGGG
12.	*Pst I*	*Providencia stuartii*	CTGCA↓G
13.	*Bgl II*	*Bacillus globigii*	A↓GATCT
14.	*Hpa II*	*Hemophilus parainfluenzae*	C↓CGG
15.	*Alu I*	*Arthrobacter luteus*	AG↓CT
16.	*Hae III*	*Hemophilus aegyptius*	GG↓CC
17.	*Hpa I*	*Hemophilus parainfluenzae*	GTT↓AAC
18.	*Mbo I*	*Moraxella bovis*	N↓GATC
19.	*NCo I*	*Nocardia corallina*	G↓CATGG
20.	*Sac I*	*Streptomyces achromogenes*	GAGCT↓C
21.	*Sal I*	*Streptomyces albus G*	G↓TCGAC
22.	*Sau 3AI*	*Staphylococcus aureus 3AI*	N↓GATC
23.	*Taq I*	*Thermus squatius YTI*	T↓CGA
24.	*Hinc II*	*Haemphilus influenzae Rc*	GTPyPuAC
25.	*Acc I*	*Acinetobacter calcoaceticus*	GT↓(AC)(GT)AC
26.	*Dde I*	*Desulfovibrio desulfuricans*	C↓GNAG
27.	*Eco RII*	*E.coli R-245*	↓CC(A/T)GG
28.	*Hinf I*	*Haemophilus influenzae Rf*	G↓ANTC
29.	*Mst II*	*Microcoleus species*	CC↓TNAGG

If cut is not shown it means exact cutting size is not known. Parentheses indicate that either bracketed base will suffice recognition.

Note : Py, Pu indicate non-specific pyrimidines or purines. N indicates any nucleotide base.

Complementary DNA (cDNA)

A second method for obtaining a gene of interest is to create a *complementary DNA (cDNA)* strand from a strand of messenger RNA (mRNA). This procedure is used when specific mRNA molecules can be isolated from cells that produce high levels of a certain protein. For example, a protein called *factor VIII* is necessary for proper blood clotting. It is the protein that most hemophiliacs are missing because

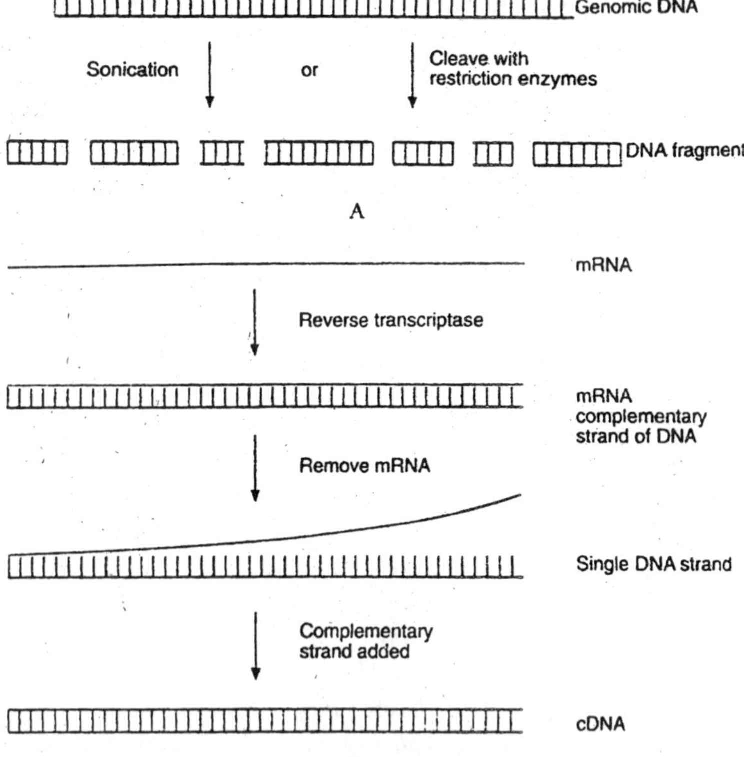

Fig. 11.5. A—To obtain DNA for cloning, large molecules are first broken into smaller segments by sonication or by cleaving them with restriction enzymes. The DNA fragments can then be separated and analyzed to determine which contain a gene of interest. B—If the mRNA encoding a gene of interest can be isolated, a DNA strand that is complementary to this mRNA can be synthesized.

of a mutation in the relevant encoding gene. From several cleaver studies, high levels of this protein were found to be produced in the liver along with the key mRNA molecules directing its synthesis

By separating the factor VIII protein and the affiliated mRNA from other proteins and other mRNAs that were also present, it was possible to identify the gene encoding factor VIII. The major advantage of this method is that a "pure" DNA sequence can be produced that does not contain the extraneous DNA sequences (introns) that occur in genomic DNA. Once the mRNA is isolated and purified, it is used as a template to construct a cDNA strand using the unique enzyme *reverse transcriptase*, which synthesizes DNA from RNA. When the cDNA has been formed, the mRNA strand is removed, and a DNA strand that is complementary to the cDNA is then added, resulting in a typical double-stranded molecule. Once DNA has been obtained, the next step in gene cloning is to create an rDNA molecule that can be used for cloning cDNA or any of the thousands of fragments that can be obtained from genomic DNA.

VECTORS

Vectors are the carrier DNAs into which 'foreign' DNAs or genes of interest are spliced to make a recombinant DNA. Vectors along with this 'foreign' DNAs (i.e., recombinant DNA) are then introduced into appropriate host cell and are maintained for study or expression.

Vectors are essentially expected to replicate inside the host cell along with the inserted DNA. The complete technique of isolating a gene of interest and inserting it into a vector and then replicating and maintaining it into a host cell is called as *Gene cloning*.

Types of Vectors

There are two types of vectors:

(a) Cloning vectors.

(b) Expression Vectors.

(a) Cloning vectors

Cloning vectors are used for obtaining millions of copies of cloned DNA segment. The cloned genes in these vectors are not expected to express themselves, at transcription or translation level. Cloning vectors are used for creating genomic library or preparing, the probes or genetic engineering experiments or other basic studies.

(b) Expression vectors

Expression vectors allow the expression of cloned gene, to give the product (protein). This can be achieved through the use of promoters and expression cassettes and regulatory genes (sequences). Expression vectors are used for transformation to generate transgenic plant, animal or microbe where cloned gene expresses to give the product (protein). Commercial production of product of cloned gene may also be achieved by high level expression using the expression vectors.

Promoters in Vectors

Routine manipulations in gene cloning experiments do not require expression of the cloned DNA. But when required, selection of host/ vector system plays important role. If cDNA molecule is used for cloning eukaryotic gene in prokaryotic system then problem of post transcriptional modification is obviated. For given host cell, there may be several types of expression vectors, then in addition of other consideration, a key feature of selection of expression vector is the type of promoter used to direct expression of the cloned sequence.

A vector with very efficient promoter is chosen to maximise the expression. Promoters are referred to as strong or weak depending on their efficiency of expression. Weak promoters are used if overexpression and excess product formation is likely to be toxic to host cell. Promoters are regions with a specific base sequence, to which RNA polymerase will bind. In addition to the strength of promoter, it is also desirable to regulate expression. This is done by using promoters that are either inducible or repressible. This exerts some control.

Arrangement of restriction sites immediately downstream from the promoter is critical. Thus expression vector will have following essential features:

(a) Origin of replication that is functional in target host cell.

(b) Antibiotic resistance genes or other genetic selection mechanism.

(c) Promoter (strong or weak as per suitability) along with regulatory control.

(d) Restriction sites immediately downstream form the promoter.

(e) Unique restriction site for cloning located in a position where the inserted cDNA sequence can be expressed effectively.

For expression of cloned gene in plants or animals only plant specific or animal specific promoters work. Promoters can confer the transferred gene a specific pattern of expression.

Expression Cassettes

These are gene constructs which allow the insertion of foreign genes, either as transcriptional or transnational fusions, behind specific promoters. PRT series of plasmids which have been derived from pUC 18/19 are the examples.

Organism	*Gene promoter*	*Induction by*
1. *E.coli*	lac operon, trp operon, λ P_L	IPTG, β-indolylacetic acid, temperature sensitive λ cl protein resp.
2. *A. nidulans*	Glucoamylase	Starch.
3. *S. cerevisiae*	Acid phosphatase Alcohol dehydrogenase galactose utilization Metallothionein	Phosphate depletion glucose depletion galactose heavy metals.
4. *T. reesei*	Cellobiohydrolase	Cellulose
5. Mouse	Metallothionein	heavy metals
6. Human	Heat shock protein	Temperature 740°C
7. Plants (in basal region)	Nopaline synthase (Nos) Mannopine synthase (Mas)	
8. Plants (in leaf tissue)	35 S promoter of Ca MV [high expression (10 fold) at distance of 2 kbp]	
9.	Polyhedrin promoter from baculovirus (used for developing biopesticides or for production of specific chemicals in industry)	

Expression Vectors

Expression vectors are simply designed to express detectable levels of foreign proteins usually at the bench-scale level (i.e. in shake flask cultures).

Production Vectors

Production level vectors are designed for stable large scale production of gene products at economically significant levels. This means production level vectors have additional genetic elements such as *par* sequences, transcription terminators downstream from the cloned genes etc. This is only artificial distinction. Efficients expression vectors are used successfully to express wide variety of proteins. Plasmids like *pASI* are attractive for production purposes as they have strong *pL* transcriptional unit and also a strong transnational unit.

Cloning in *E.coli*

Plasmid DNA used as vector can be cleaved at a site with restriction endonuclease enzyme and foreign DNA segments can be inserted here. Generally, plasmids with two genes for different antibiotic resistance are used—one to identify bacteria that carry plasmid and second to distinguish chimeric plasmid from the parental vector. Insertion of foreign DNA segment is done in site at one antibiotic resistance gene, therefore resistance is lost to one antibiotic.

The chimeric plasmid is resistant to only one antibiotic while the parental plasmid is resistant to two antibiotics. Multicopy plasmids are more suitable since 10-30 copies of one plasmid in a cell along with the inserted gene will give an amplified response. Addition of chloramphenicol often affects new rounds of chromosome replication while plasmid replication is unaffected. This makes isolation of plasmid easier. A common method for cloning DNA is to cleave both plasmid and insert DNA with the same restriction endonuclease. Most of the plasmids which are used as cloning vectors are plasmids derived from natural plasmids so as to posses desirable features.

pBR 322

This is derived from *E. coli* plasmid Col E1. Bolivar and Rodriguez prepared this vector hence the name and 322 are numericals which were significant to these scientists. Alternations and restructuring is done while deriving these plasmids from original natural plasmids. Genes for relaxed replication and genes for antibiotic resistance are inserted in them. pBR 322 is 4362 bp long DNA with genes for resistance to tetracycline and ampicillin. pBR 322 possesses on origin of replication and restriction sites for cleavage by variety of restriction enzymes.

Characteristics of pBR 322

1. It is much smaller in size than natural plasmid. Advantages of small size are : (i) easier to handle, (ii) less susceptible to physical damage, (iii) simpler restriction man, (iv) easy uptake by bacteria during transformation, (v) higher copy number therefore easy detection.
2. Bacterial origin of replication is present and this ensures that plasmid will be replicated in host.
3. Origin of replication is relaxed here, hence activity is not tightly linked to cell division. Resultantly plasmid replication will be initiated for more frequently than chromosomal replication. So more copy number of plasmid per cell is possible.
4. Two genes for different antibiotic resistance.
5. There are single restriction sites for number of enzymes, scattered on plasmid and can be used for cleavage and insertion. Presence

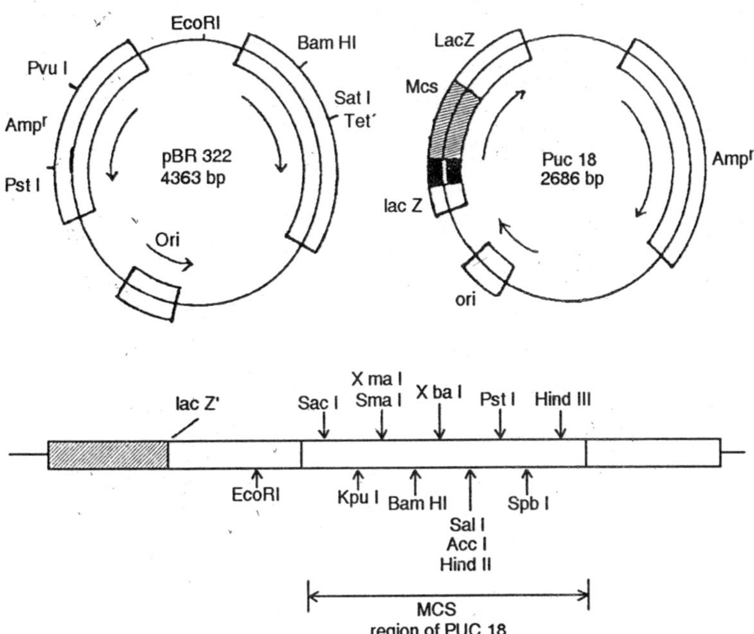

Fig. 11.6. Plasmids pBR 322 and pUC 18.

of such sites within antibiotic resistance gene are important, as they cause respective loss of resistance.

pAT 153 and *pXf 3* are two derivatives of pBR 322 having smaller size and higher copy number.

pBR 327

This is derived from pBR 322, by deletion of nucleotides between 1427 and 2516. These nucleotides are deleted to reduce size of vector and reduce interference. pBR 327 still has genes for resistance against tetracycline and ampicillin.

pUC Vectors

The name is derived from the place of their initial preparation (i.e., University of California). These vectors have 2700 base pairs and ampicillin resistance gene and lac Z gene. Insertion of foreign DNA in lac Z gene causes inactivation of lac Z gene. *E. coli* which is lactose –ve is used for transformation. When culture is grown in presence of IPTG (Isopropyl thiogalactoside) which induces the synthesis of β galactosidase) and X-gal substrate, *E. coli* (transformed with pUC vector with insert) gives white colonies while *E. coli* (transformed with pUC vector but without insert) gives blue colonies. (Substrate is chromogenic hence coloured product is formed if enzyme is active). Vectors of pUC family have a region that contains several unique restriction sites in a short stretch of DNA. This region is known as a polylinker or multiple cloning site (MCS).

Bacteriophages Vectors

Phage has a linear DNA molecule so a single break creates two fragments. Foreign DNA can be inserted between them and two fragments can be joined. Such phages when undergo lytic cycle in host will produce more chimeric DNA. Wild type λ phage could accommodate only 2.5 kb of foreign DNA. Phage vectors are restructured by removing nonessential genes and making vector DNA smaller so that larger insert can be accommodated in phage head during packing. Lambda phage (λ) such prepared has one Eco Rl site and accommodates 20-25 kb of foreign DNA. They are used for preparing genomic library of eukaryotes. One other λ phage devised has two Bam HI sites that flank the I/E region.

Insertion vectors

Insertion phage vector has single restriction site which is used for insertion of foreign DNA. Smaller foreign DNA can be packed here.

Replacement vectors

Replacement vector has two restriction sites which flank a region known as stuffer fragment. Larger fragment of foreign DNA can be replaced between these two sites.

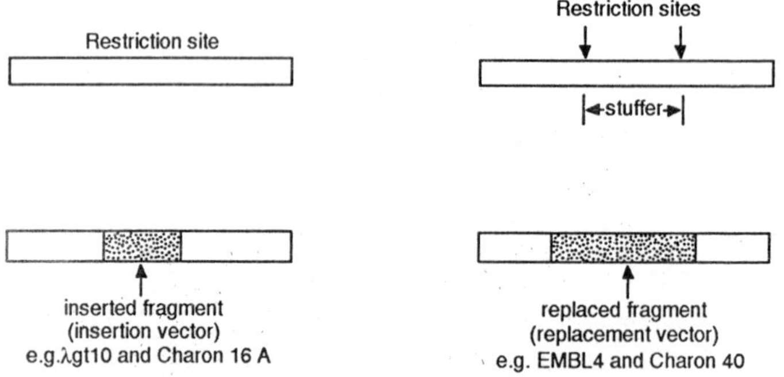

Fig. 11.7. Insertion and replacement vectors.

Cosmids

Cosmids are the novel cloning vectors which possess properties of both plasmid and λ phage. Cosmids first were developed in 1978 by

Barbara Hohn and John Collins. Cosmids contain a cos site of λ phage (which is essential for packaging of nucleic acid into protein coat) plus essential features of plasmid and several unique restriction sites for insertion of DNA to be cloned. Cosmids can be perpetuated in bacteria in plasmid form, but can be purified by packaging *in vitro* into phages. When viral DNA is injected into the cell as a linear molecule it's both ends are cohesive and complementary to each other, 12 base in length.

Once inside the cell, these ends base pair and become permanently joined by ligation to form a region which is known as *cos site* and give rise to circular DNA molecule. Replication of DNA by rolling circle mechanism results into a concatemer which is a long molecule made up of many copies of viral DNA linked end to end through cos sites. DNA is packaged by looping regions between cos sites into precursor of the viral head. When head is full, the cos sites should be at the mouth of the head, where they will be cleaved to generate a linear molecule with cohesive ends. Subsequently tail proteins are added to give infective particle. Thus only requirement for length of DNA to be packaged into viral heads is that it should contain cos sites spaced at correct distance. This distance ranges from 37 to 52 kb. Cos sites of λ possess 280 bp flanking sequence.

For cloning foreign DNA into cosmid vector, cosmid DNA is first linearised by cutting it with appropriate restriction enzyme. Then it is treated with the calf intestinal phosphatase to remove phosphate groups (5′) at its ends so as to prevent recircularisation of cosmid DNA. Foreign DNA which is to be cloned is also treated with the same restriction enzyme which was used for treating cosmid DNA. Subsequently cosmid DNA and foreign DNA fragments are mixed in presence of T4 DNA ligase. Many different products generate in the mixture.

A product may form where foreign DNA binds to cos site of cosmid in same orientation. This on *in vitro* incubation with phage head and tail proteins, cleavage of cos sites takes place and intervening DNA is packed into phage particles. When susceptible bacteria are infected by such phage, hybrid DNA is injected into cell. Then it functions like plasmid and expresses antibiotic resistance gene. The infected cells can be selected and isolated on basis of antibiotic resistance gene expressed. Foreign DNA presence can be detected in antibiotic resistant bacterial colonies. Inserts of 40-50 kb length can be easily placed between cos sites and to get packageble forms of cosmids.

Advantage of using cosmid vector is that larger DNA can be cloned than what is possible with phage or plasmid. As larger inserts are possible genomic library can be created which is composed of fewer clones to be screened. Efficiency of cosmids is high enough to produce a complete genomic library of 10^6–10^7 clones from a mere 1μg of insert. Genomic libraries of Drosophila, mouse and several other organisations has been produced with cosmid vectors. Cosmid PLFR-5 has two cos sites, 6 restriction enzymes target sites, origin of replication and tetracycline resistance marker. It can insert 50 kb of DNA. Cosmid system in case of P_1 bacteriophage of *E. coli* can carry 85 kb of inserted DNA.

For *in vitro* packaging of recombinant DNA containing cosmid *packaging extract* is used. This is nothing but two strains of bacteria. These two strains when mixed with concatameric recombinant DNA under suitable conditions head, tail and DNA with proper cos site distance is available and phage particles are produced. These infective phages then can be used to infect *E. coli* cells.

Phasmid Vectors

Here combination of plasmid and λ phage is produced. Plasmid is inserted into phage λ genome by means of site specific recombination mechanism of the phage that is normally used by phage for insertion into bacterial chromosome during lysogen formation. This process is called 'lifting' the plasmid and combination is called phasmid.

Phasmid contain functional origins of replications of the plasmids and of λ and may be propagated as plasmid or phage in appropriate *E. coli* host strains. Plasmids can be released by several of lifting. Phage particles are easy to store, have infinitive shelf life and their screening as plaques by hybridization gives better results than screening of bacterial colonies. So plasmid with cloned gene can be lifted by phages and conveniently handled. Release of recombinant plasmid is easier. λ ZAP is highly developed phasmid containing λ, M_{13} and T_7 phages. λ ZAP is suitable for cloning cDNAs. λ ZAP has multiple unique cloning sites, can hold 10 kb of inserts, easy excision of cloned DNA possible. Insertional inactivation of β galactosidase gives blue/white screening on X gal plates. Expression of hybrid polypeptides is analogous to that in λ gt 11.

Shuttle Vectors

Transfer of genes between unrelated species is one of the requirements of molecular biotechnology. Broad host range vectors exist in Gram negative bacteria and *Streptomyces* naturally. A shuttle vector however may be required having necessary replicon for maintenance in different combinations of unrelated hosts. Shuttle vectors have potential importance in the genetic manipulations of industrially important species. Shuttle vectors can exploit gene manipulative procedures of different hosts for example when *E. coli* amplification will be possible. Shuttle vectors exist for *E. coli* yeast cells combination. *E. coli—Agrobacterium* combination and *E. coli—B. subtilis*. *E. coli—Streptomyces lividans* and *E. coli—mammalian* cells combinations.

VEHICLES

A DNA molecule needs to display several features to be able to act as a vehicle for gene cloning. Most important, it must be able to replicate within the host cell, so that numerous copies of the recombinant DNA molecule can be produced and passed to the daughter cells. A cloning vehicle also needs to be relatively small, ideally less than 10 kilobases (kb) in size, as large molecules tend to break down during purification, and are also more difficult to manipulate. Two kinds of DNA molecule that satisfy these criteria can be found in bacterial cells: plasmids and bacteriophage chromosomes. Although plasmids are frequently employed as cloning vehicles, two of the most important types of vector in use today are derived from bacteriophages.

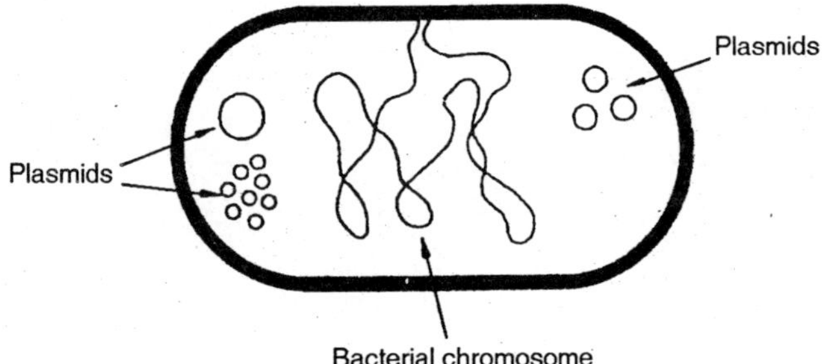

Fig. 11.8. Plasmids independent genetic elements found in bacterial cells.

Plasmids

Basic features of plasmids

Plasmids are circular molecules of DNA that lead an independent existence in the bacterial cell. Plasmids almost always carry one or more genes, and often these genes are responsible for a useful characteristic displayed by the host bacterium. For example, the ability to survive in normally toxic concentrations of antibiotics such as

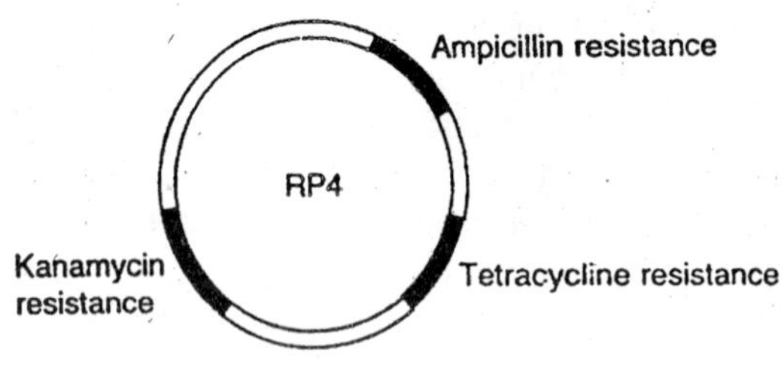

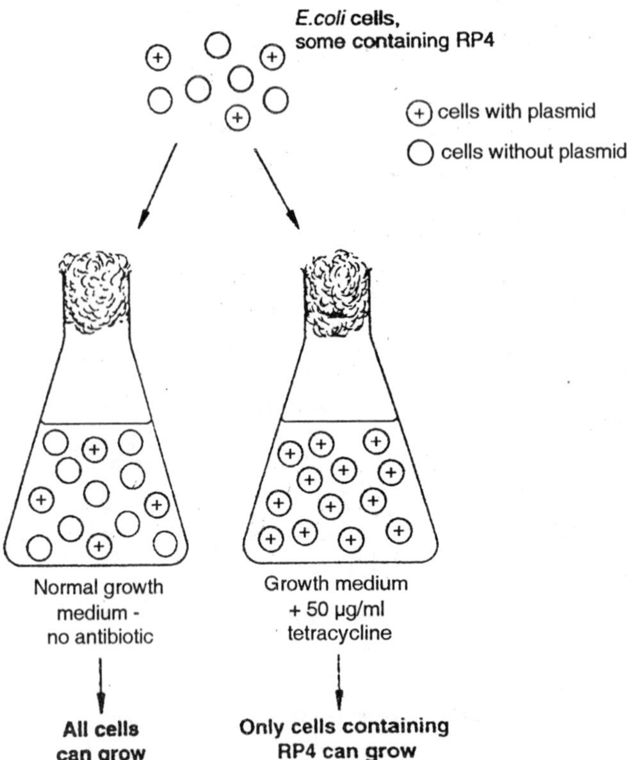

Fig. 11.9. The use of antibiotic resistance as a selectable marker for a plasmid.

chloramphenicol or ampicillin is often due to the presence in the bacterium of a plasmid carrying antibiotic resistance genes. In the laboratory antibiotic resistance is often used as a *selectable marker* to ensure that bacteria in a culture contain a particular plasmid.

All plasmids possess at least one DNA sequence that can act as an *origin of replication*, so they are able to multiply within the cell

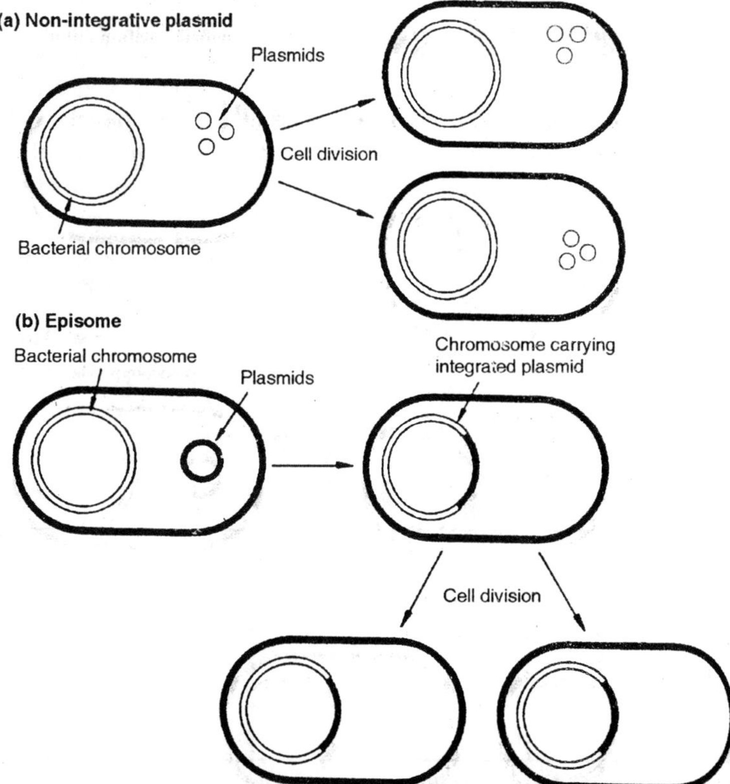

Fig. 11.10. Replication strategies for (a) a non-integrative plasmid, and (b) an episome.

quite independently of the main bacterial chromosome. The smaller plasmids make use of the host cell's own DNA replicative enzymes in order to make copies of themselves, whereas some of the larger ones carry genes that code for special enzymes that are specific for plasmid replication. A few types of plasmid are also able to replicate by inserting themselves into the bacterial chromosome. These integrative plasmids or *episomes* may be stably maintained in this form through numerous cell divisions, but will at same stage exist as independent elements. Integration is also an important feature of some bacteriophage chromosomes and will be described in more detail when these are considered.

Size and copy number

These two features of plasmid are particularly important as far as cloning is concerned. We have already mentioned the relevance of plasmid size and stated that less than 10 kb is desirable for a cloning vehicle. Plasmids range from about 1.0 kb for the smallest to over 250 kb for the largest plasmids, so only a few will be useful for cloning purposes. However, larger plasmids may be adapted for cloning under some circumstances. The *copy number* refers to the number of molecules of an individual plasmid that are normally found in a single bacterial cell.

The factors that control copy number are not well understood, but each plasmid has a characteristic value that may be as low as one (especially for the large molecules) or as many as 50 or more. Generally speaking, a useful cloning vehicle needs to be present in the cell in multiple copies so that large quantities of the recombinant DNA molecule can be obtained.

Table 11.5. Sizes of representative plasmids

Plasmid	*Size*		*Organism*
	Nucleotide length (kb)	*Molecular wt (MDa)*	
pUC8	2.1	1.8	*E.coli*
ColEI	6.4	4.2	*E.coli*
RP4	54	36	*Pseudomonas* + others
F	95	63	*E.coli*
TOL	117	78	*Pseudomonas putida*
pTiAch5	213	142	*Agrobacterium tumefaciens*

Conjugation and compatibility

Plasmids fall into two groups conjugative and non-conjugative. Conjugative plasmids are characterized by the ability to promote sexual *conjugation* between bacterial cell, a process that can result in a conjugative plasmid spreading from one cell to all the other cells in a bacterial culture. Conjugation and plasmid transfer are controlled by a set of transfer or *tra* genes, which are present on conjugative plasmids but absent from the non-conjugative type.

A non-conjugative plasmid may, under some circumstance, be cotransferred along with a conjugative plasmid when both are present in the same cell. Several different kinds of plasmid may be found in a single cell, including more than one different conjugative plasmid at any one time. In fact, cells of *E. coli* have been known to contain up to seven different plasmids at once. To be able to coexist in the same cell, different plasmids must be *compatible*. If two plasmids are incompatible then one or the other will be quite rapidly lost from the

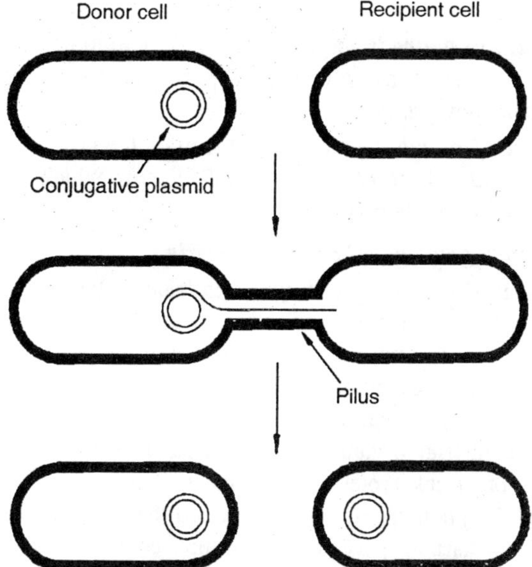

Fig. 11.11. Plasmid transfer by conjugation between bacterial cells.

cell. Different types of plasmid can therefore be assigned to different *incompatibility groups* on the basis of whether or not they can coexist, and plasmids from a single incompatibility groups are often related to each other in various ways. The basis of incompatibility is not well understood, but events during plasmid replication are thought to underlie the phenomenon.

Plasmid classification

There are five main types of plasmid which are as follows:

1. *Fertility* or *'F' plasmids* carry only *tra* genes and have no characteristic beyond the ability to promote conjugal transfer of plasmids, e.g., F plasmid of *E. coli*.
2. *Resistance* or *'R' plasmids* carry genes conferring on the host bacterium resistance to one or more antibacterial agents, such as chloramphenicol, ampicillin and mercury, R plasmids are very important in clinical microbiology as their spread through natural populations can have profound consequences in the treatment of bacterial infections; e.g., RP4, commonly found in *Pseudomonas*, but also occurring in many other bacteria.
3. *Col plasmids* code for colicins–proteins that kill other bacteria; e.g., ColE1 of *E. coli*.

4. *Degradative plasmids* allow the host bacterium to metabolize unusual molecules such as toluene and salicylic acid; e.g., TOL of *Pseudomonas putida*.
5. *Virulence plasmids* confer pathogenicity on the host bacterium; e.g., *Ti plasmids* of *Agrobacterium tumefaciens*, which induce crown gall disease on dicotyledonous plants.

Plasmids in organisms other than bacteria

Although plasmids are widespread in bacteria they are by no means so common in other organisms. The best characterized eukaryotic plasmid is the *2μm circle* that occurs in many strains of the yeast *Saccharomyces cerevisiae*. The discovery of the 2 μm plasmid was very fortuitous as it has allowed the construction of vectors for cloning genes with this very important industrial organism as the host. However, the search for plasmids in other eukaryotes (e.g., filamentous fungi, plants and animals) has proved disappointing and it is suspected that many higher organisms simply do not harbour plasmids within their cells.

Bacteriophages

Basic features of bacteriophages

Bacteriophages, or phages as they are commonly known, are viruses that specifically infect bacteria. Like all viruses, phages are very simple in structure, consisting merely of a DNA (or occasionally RNA) molecule carrying a number of genes, including several for replication of the phage, surrounded by a protective coat or *capsid* made up of protein molecules.

The general pattern of infection, which is the same for all types of phage, is a three-step process.

1. The phage particle attaches to the outside of the bacterium and injects its DNA chromosome into the cell.
2. The phage DNA molecule is replicated, usually by specific phage enzymes coded by genes on the phage chromosome.
3. Other phage genes direct synthesis of the protein components of the capsid, and new phage particles are assembled and released from the bacterium.

With some phage types the entire infection cycle is completed very quickly, possibly in less than 20 minutes. This type of rapid infection is called a *lytic cycle*, as release of the new phage particles is associated with lysis of the bacterial cell. The characteristic feature of a lytic infection cycle is that phage DNA replication is immediately

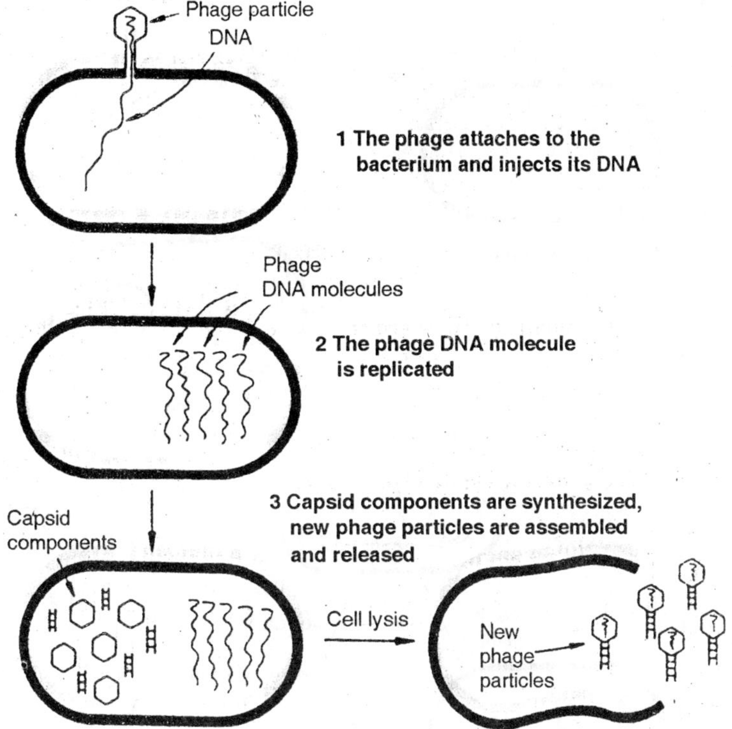

Fig. 11.12. The general pattern of infection of a bacterial cell by a bacteriophage.

followed by synthesis of capsid proteins, and the phage DNA molecule is never maintained in a stable condition in the host cell.

Lysogenic phages

In contrast to a lytic cycle, *lysogenic* infection is characterized by retention of the phage DNA molecule in the host bacterium, possibly for many thousands of cell divisions. With many lysogenic phages the phage DNA is inserted into the bacterial genome, in a manner similar to episomal insertion. The integrated form of the phage DNA (called the *prophage*) is quiescent, and a bacterium (referred to as a *lysogen*) which carries a prophage is usually physiologically indistinguishable from an uninfected cell.

The prophage is eventually released from the host genome and the phage reverts to the lytic mode and lyses the cell. The infection cycle of λ, a typical lysogenic phage is of this type. A limited number of lysogenic phages follow a rather different infection cycle. When M13, or a related phage, infects *E. coli*, new phage particles are continuously

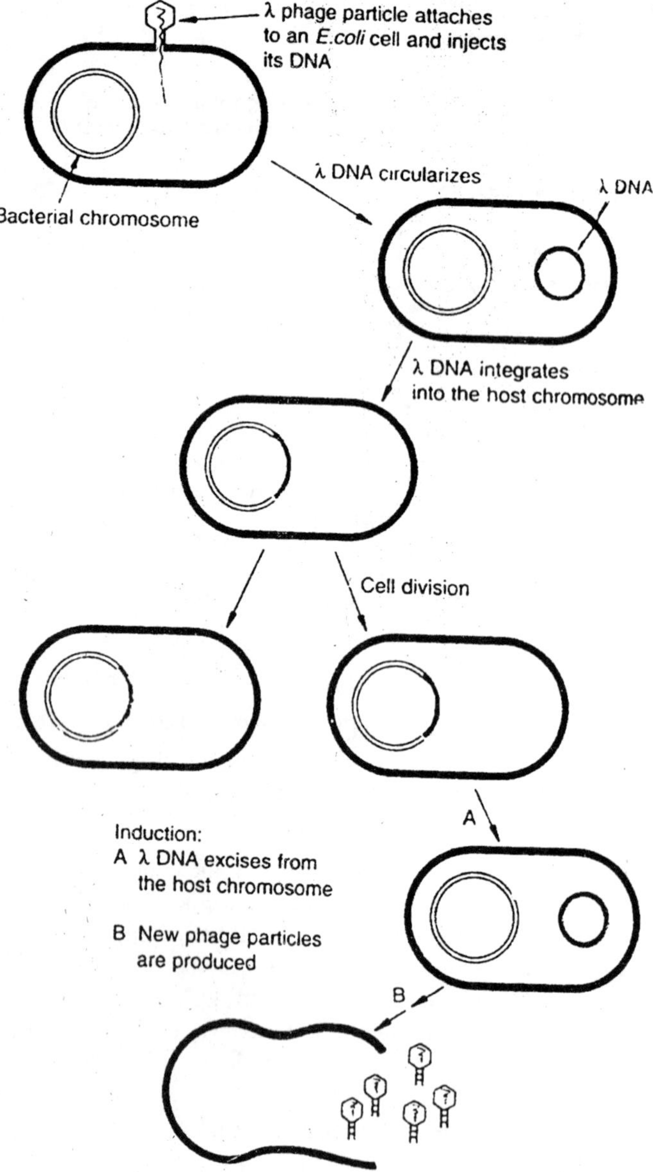

Fig. 11.13. The lysogenic infection cycle of bacteriophage λ.

assembled and released from the cell. The M13 DNA is not integrated into the bacterial genome and does not become quiescent. With these

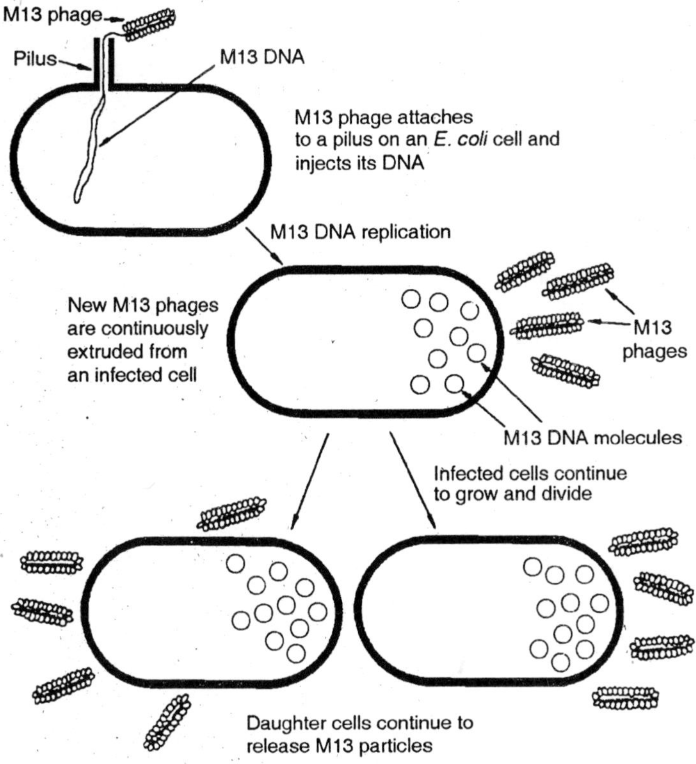

Fig. 11.14. The infection cycle of bacteriophage M13.

phages, cell lysis never occurs, and the infected bacterium can continue to grow and divide, albeit at a slower rate than uninfected cells. Although there are many different varieties of bacteriophage, only λ and M13 have found any real role as cloning vectors. The properties of these two phages will now be considered in more detail.

(a) Gene organization in the λ DNA molecule

λ is a typical example of a head-and-tail phage. The DNA is contained in the polyhedral head structure and the tail serves to attach the phage to the bacterial surface and to inject the DNA into the cell. The λ DNA molecule is 49 kb in size and has been intensively studied by the techniques of gene mapping and *DNA sequencing*. As a result the positions and identities of most of the genes on the l DNA molecule are known.

A feature of the λ genetic map is that genes related in terms of function are clustered together on the genome. For example, all of the genes coding for components of the capsid are grouped together in the left-hand third of the molecule, and genes controlling integration of the prophage into the host genome are clustered in the middle of the molecule. Clustering of related genes is profoundly important for controlling expression of the λ genome, as it allows genes to be switched on and off as a group rather than individually. Clustering is also important in the construction of λ-based cloning vectors.

(b) The linear and circular forms of λ DNA

A second feature of λ that turns out to be of importance in the construction of cloning vectors is the conformation of the DNA molecule. The molecule is linear, with two free, ends, and represents the DNA present in the phage head structure. This linear molecule consists of two *complementary* strands of DNA, base-paired according to the *Watson-Crick rules* (that is, double-stranded DNA). However, at either end of the molecule is a short 12-nucleotide stretch, in which the DNA is single-stranded.

The two single strands are complementary, and so can base-pair with one another to form a circular, completely double-stranded molecule. Complementary single strands are often referred to as *'sticky' end* or 'cohesive' ends, because base-pairing between them can 'stick' together the two ends of a DNA molecule (or the ends of two different DNA molecules). The λ cohesive ends are called he *cos sites* and they play two distinct roles during the λ infection cycle.

First of all, they allow the linear DNA molecule that is injected into the cell to be circularized, which is a necessary prerequisite for insertion into the bacterial genome. The second role of the *cos* sites is rather different, and comes into play after the prophage has excised from the host genome. At this stage a large number of new λ DNA molecules are produced by the rolling circle mechanism of replication in which a continuous DNA strand is 'rolled off' of the template molecule. The result is a catenane consisting of a series of linear λ genomes joined together as the *cos* sites. The role of the *cos* sites is now to act as recognition sequences for an *endonuclease* which cleaves the catenane at the *cos* sites producing individual λ genomes.

The endonuclease (which is the product of gene A on the λ DNA molecule) creates the single-stranded sticky ends, and also acts in conjunction with other proteins to package each λ genome into a phage head structure. The cleavage and packaging processes recognize just the *cos* sites and the DNA sequences to either side of them. Changing the structure of the internal regions of the λ genome, for example by inserting new genes, has no effect on these events so long as the overall length of the λ genome is not altered too greatly.

(c) M13—a filamentous phage

M13 is an example of a filamentous phage and is completely different in structure from λ. Furthermore, the M13 DNA molecule is much smaller than the λ genome, being only 6407 nucleotides in length. It is circular, and is unusual in that it consists entirely of single-stranded DNA. The smaller size of the M13 DNA molecule means that it has room for fewer genes than the λ genome. This is possible because the M13 capsid is constructed from multiple copies of just three proteins, whereas synthesis of the λ head-and-tail structure involves over 15 different proteins. In addition, M13 follows a simpler infection cycle than λ and does not need genes for insertion into the host genome. Injection of an M13 DNA molecule into an *E. coli* cell occurs via the *pilus*, the structure that connects two cells during sexual conjugation.

Once inside the cell the single-stranded molecule acts as the template for synthesis of a complementary strand, resulting in normal double-stranded DNA. This molecule is not inserted into the bacterial genome, but instead replicates until over 100 copies are present in the cell. When the bacterium divides, each daughter receives copies of the phage genome, which continues to replicate, thereby maintaining its overall numbers per cell. New phage particles are continuously assembled and released, about 1000 new phages being produced during each generation of an infected cell.

(d) The attraction of M13 as a cloning vehicle

Several features of M13 make this phage attractive as the basis for a cloning vehicle. The genome is less than 10 kb in size, well within the range that we stated was desirable for a potential vector. In addition, the double-stranded *replicative form* of the M13 genome behaves very much like a plasmid and can be treated as such for experimental purposes. It is easily prepared from a culture of infected *E. coli* cells and can be reintroduced by *transfection*.

Most importantly, genes cloned with an M13-based vector can be obtained in the form of single-stranded DNA. Single-stranded versions of cloned genes are useful for several techniques, notably DNA sequencing and *in vitro* mutagenesis. Using an M13 vector is an easy and reliable way of obtaining single-stranded DNA for this type of work.

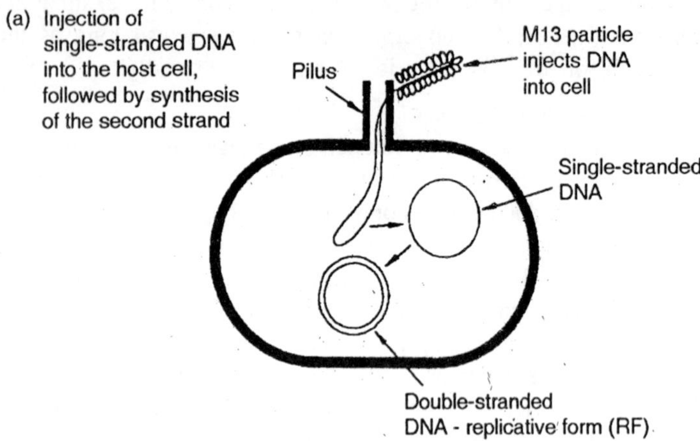

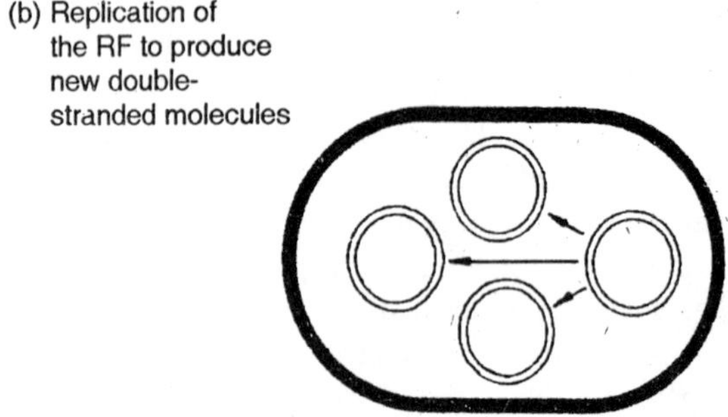

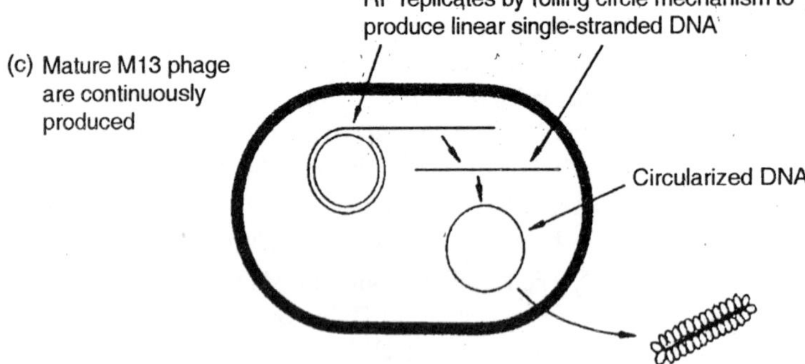

Fig. 11.15. The M13 infection cycle showing the different types of DNA replication.

Viruses as cloning vehicles for other organisms

Most living organisms are infected by viruses and it is not surprising that great interest has been shown in the possibility that viruses might be used as cloning vehicles for higher organisms. This is especially important when it is remembered that plasmids are not commonly found in organisms other than bacteria and yeast. In fact viruses have considerable potential as cloning vehicles for animal cells. Mammalian viruses such as *simian virus 40 (SV40)* and *adenoviruses*, and the insect *baculoviruses*, are the ones that have received most attention so far, but others are also being studied.

The Role of Bacteria

Bacteria are one-celled organisms that reproduce very rapidly by binary fission (dividing into two cells) after a brief growth period. They live in practically every environment, but the species of greatest use to molecular biologists are those that can be easily cultivated in the laboratory by placing them on a suitable culture medium contained in a small dish. A single bacterial cell can give rise to tens or hundreds of thousands of identical cells (clones) in a single day. When a vector containing rDNA is inserted successfully into a bacterial cell, it is replicated, and hence the DNA fragment of interest is cloned.

Much of the earliest information about cell biology and molecular biology came from studies of bacteria. Today, they play a critical role in gene-cloning technologies. The most popular bacterium in all of this work has been *E. coli*, a normally harmless inhabitant of mammalian digestive tracts. More is known about the molecular biology of *E. coli* than about any other species on Earth. *E. coli* and some other bacteria are used to perform two functions in rDNA technology. First, they are used to clone genomic DNA or cDNA to obtain unlimited numbers of genes that can be used in further studies. Second, pure cultures, in which every bacterium contains a gene encoding a desired protein, can be created and put to work manufacturing huge quantities of that protein. Plasmids that contain specific sequences to direct protein synthesis are called *expression vectors*.

Gene Libraries

One of the major uses of rDNA technology is to create *gene libraries* for different species. In accordance with the procedure used to obtain the DNA, there are libraries of both genomic clones and cDNA clones. The number of "volumes" (DNA fragments or genes) required in the library to ensure that all gene sequences are represented is generally related to the size and complexity of the species. For *E. coli*, 1,500 fragments are necessary; for yeast, 4,600; fruit flies, 48,000; and humans, 800,000.

Researchers create a *gene library* for a particular species in the following sequence of events:

1. Genomic DNA is cut into thousands of fragments with appropriate restriction enzymes.
2. Each fragment, representing approximately one gene, is spliced into a phage-cloning vector.
3. The phages are replicated by the "host" bacteria, thus replicating (cloning) the inserted gene fragments.

Such a library is a repository of the entire genome where every gene of a species is represented. The space required for a typical gene library is a test tube or a set of small culture dishes that can be conveniently kept in a refrigerator pending further analysis. Molecular biologists have developed specialized techniques to "screen" gene libraries, that is, to determine where specific genes are located in the pieces of DNA. This involves isolating a gene and identifying the protein it encodes. Once a gene has been isolated and cloned, it is possible quickly to determine its entire base sequence. As you might expect, the greatest emphasis is currently on research related to the human gene library, and remarkable progress has been made in isolating human genes of medical importance. Nevertheless, the human gene library is so immense, and many genes are so difficult to locate, that a complete reading of our library probably lies one or more decades in the future.

Applications of Genetic Engineering

Recombinant DNA technology has revolutionized biology in the past decade. At present, its main uses are (1) facilitating the production of useful proteins, (2) creating bacteria capable of synthesizing economically important molecules, (3) supplying DNA and RNA sequences as a research tool, (4) altering the genotype of organisms such as plants, and (5) potentially correcting genetic defects in animals (gene therapy). Some examples of these applications follows:

Commercial Possibilities

Bacteria with novel phenotypes can be produced by genetic engineering sometimes by combining the features of several other bacteria. For example, several genes from different bacteria have been inserted into a single plasmid that has then been placed in a marine bacterium, yielding an organism capable of metabolizing petroleum; this organism has been used to clean up oil spills in the oceans.

Many biotechnology companies are at work designing bacteria that can synthesize industrially important chemicals. Bacteria have been designed that are able to compost waste more efficiently and to fix nitrogen (to improve the fertility of soil), and an enormous effort is currently being expended to create organisms that can convert biological waste to alcohol. A human insulin, synthesized in *E. coli*, is already commercially available. Altering the genotype of plants is an important application of recombinant DNA technology. Of great use is the bacterium *Agrobacterium tumefaciens* and its plasmid Ti, which produces crown gall tumors in dicotyledonous plants. These tumor result from disruption of the bacterium inside plant cells, release of bacterial DNA, and integration of a segment of the plasmid DNA into the plant chromosome.

It is possible by genetic engineering to introduce genes from one plant into this plasmid and then, by infecting a second plant with the bacterium, transfer the genes of the first plant to the second plant. (Actually genes are first cloned in an *E. coli* plasmid and then recloned in Ti). Attempts are being made to perform plant breeding in this way. An example is the attempted alteration of the surface structure of the roots of grains such as wheat, by introducing certain genes from legumes (peas, beans), in order to give grains the ability of the legumes to establish root nodules of nitrogen-fixing bacteria. If successful, this would eliminate the need for the addition of nitrogenous fertilizers to grain-growing soils.

The first engineered recombinant plant of commercial value was developed in 1985. An economically important herbicide (weed killer) is glyphosate, which inhibits a particular essential enzyme in many plants. However, most herbicides cannot be applied to fields growing crops because both the crop and the weeds would be killed.

The target gene of glyphosate is also present in the bacterium *Salmonella typhimurium*. A resistant form of the gene was obtained by mutagenesis and growth of *Salmonella* in the presence of glyphosate; the gene was cloned in *E. coli*, and then recloned in *Agrobacterium*. Infection of plants with purified Ti containing the glyphosate-resistance gene has yielded varieties of maize, cotton, and tobacco that are resistant to glyphosate. Thus, fields of these crops can be sprayed with glyphosate at any stage of growth of the crop. All weeds are killed, and the crop is totally unharmed. An interesting bacterium is a strain of *Pseudomonas fluorescens*, which lives in association with maize and soybean roots. A lethal gene from *Bacillus thuriengis*, a bacterium pathogenic to the back cutworm, has been engineered into this bacterium. The black cutworm causes extensive crop damage and is usually combatted with noxious insecticides. In preliminary studies inoculation of soil with the engineered *Ps. fluorescens* resulted in death of the cutworm.

Uses in Research

Recombinant DNA technology is extraordinarily useful in basic research. Several examples were already given in mutant bacteria in which particular genetic systems have been altered in an effort to make them more amenable to study (for example, the numerous mutants in the *lac* operon). Such mutants have usually been derived through standard genetic techniques. For simple mutants this procedure is straightforward, but for mutants required to have many genetic markers (which may be very closely linked) the frequency of production of mutants can be so low that their isolation becomes very tedious.

Recombinant DNA techniques can simplify mutant construction, since fragments containing desired genetic markers can be purified, altered, and combined in test tube, and then introduced into another cell. This saves time and labour, and often enables mutants to be constructed that cannot in practice be formed in any other way. An example is the formation of double mutants of animal viruses, which undergo crossing over at such a low frequency that mutations can rarely be recombined by genetic crosses.

The greatest impact of the new technology on basic research has been in the study of eukaryotes, in particular, of eukaryotic regulation. Experiments of the type, which study the regulation of operons in bacteria, have been made possible by the use of mutations in promoters, operators, and structural genes. However, this approach has not been feasible with eukaryotes, because eukaryotes are diploid and, hence, mutants are difficult to isolate.

Furthermore, except for the unicellular eukaryotes such as yeast, there is no simple and rapid way to do multiple genetic manipulations with eukaryotic cells. Cloning techniques have made it possible to study regulation of gene expression in eukaryotes by direct assays of mRNA molecules produced by particular genes. The approach is based on the success with which it has become possible to understand gene activity in bacteria by studying the synthesis of mRNA in a variety of conditions.

An excellent assay is DNA-RNA hybridization, in which either a primary transcript or mRNA is detected by hybridization to a DNA sample that has been enriched for the gene being studied. In microbial experiments specialized transducing particles would be the source of the DNA—for example, *E. coli* gal mRNA is assayed by hybridization of radioactivity labeled intracellular RNA with DNA of *l gal* transducing particles. However, transducing particles do not exist for eukaryotic systems, so recombinant DNA techniques have been used to clone a gene whose regulation is to be studied.

The usual procedure is to clone the gene first in a plasmid (or a phage) vector and then allow the cell containing the plasmid to multiply. Purified plasmid DNA provides the researcher with a large supply of the DNA of that gene. The vector containing that DNA sequence is called a DNA *probe*, because its DNA, in denatured form, can be added to a cell extract containing mRNA, to probe for a particular mRNA by renaturation. Since no natural genes of the vector are present in eukaryotic cells, the vector DNA can be regarded as a source of pure DNA copies of a particular eukaryotic gene because no other genes in the vector will participate in the renaturation.

A further useful technique is to apply denatured probe DNA to a nitrocellulose filter and then to use a filter-binding assay for the specific RNA species. DNA but not RNA binds to these filters. Radioactive mRNA can then be incubated with a filter containing the DNA, using conditions such that renaturation will result. Except where the mRNA has renatured to the denatured DNA, the RNA can be removed by washing the filter. Thus, presence of radioactivity in the filter after washing indicates that complementary mRNA has been bound to the DNA on the filter. This simple technique and variations such as Southern blotting have revolutionized the study of eukaryotic gene regulation.

Production of Eukaryotic Proteins

One of the most valuable applications of genetic engineering is the production of large quantities of particular proteins that are otherwise difficult to obtain. The method is simple in principle. The gene encoding the desired protein is cloned in a vector adjacent to a bacterial promoter, and tests are performed to ensure that the gene is oriented such that its coding strand is linked to the strand containing the promoter. A plasmid for which many copies are present in each cell, or occasionally an actively replicating phage (such as λ), is used as a vector; in both cases, cells can be prepared that contain several hundred copies of the gene, and this can result in synthesis of a gene product to reach a concentration of about 1 to 5 percent of the cellular protein. In practice, production of large quantities of a prokaryotic protein in a bacterium is straightforward. However, if the gene is from a eukaryote, special problems exist:

1. Eukaryotic promoters are not usually recognized by bacterial RNA polymerases.
2. The mRNA transcribed from eukaryotic genes may not be translatable on bacterial ribosomes.
3. Introns may be present, and bacteria are unable to excise eukaryotic introns.
4. The protein itself often must be processed (for example, insulin); and bacteria cannot recognize processing signals from eukaryotes.
5. Eukaryotic proteins are often recognized as foreign material by bacterial protein-digesting enzyme and are broken down.

Several approaches to solving these problems have been taken with some degree of success. One procedure uses a genetically engineered plasmid, called a *shuttle vector*, that consists of both *E. coli* and yeast DNA and replicates in both organisms.

Cloning is done in *E. coli*, which for a variety of technical reasons is easier than cloning in yeast, and then the plasmid is isolated and transferred to yeast for expression of the cloned gene. In yeast the five problems just stated are avoided, or at least substantially reduced. Another approach uses a plasmid containing the *E. coli lac* region, cleaved in the *lacZ* gene by a restriction enzyme making blunt ends. Either c-DNA or a synthetic DNA molecule whose sequence is known from the amino acid sequence of the protein product is inserted into the *lacZ* gene, so that the eukaryotic protein is synthesized as the terminal region of β-galactosidase, from which it can be cleaved.

The first example of this approach resulted in a synthetic gene capable of yielding a 14-residue polypeptide hormone--somatostatin— which is synthesized *in vivo* in the mammalian hypothalamus. The procedure, applicable to most short polypeptides, was the following.

By chemical techniques, a double-stranded DNA molecule was synthesized containing 51 base pair; the base sequence of the coding strand was

TAC-(42 base encoding somatostatin)-ACTATC

The base sequence of the corresponding mRNA molecule was

AUG-(42 bases encoding somatostatin)-UGAUAG

The AUG codon of the mRNA was not used for initiation, but specified methionine. The mRNA terminated with two stop codons UGA and UAG.

The vector was the plasmid pBR322 modified to contain the *lac* promoter-operator region and a portion of the *lacZ* gene encoding the amino-terminal segment of β-galactosidase. The vector was cleaved at a site in the *lacZ* segment by a restriction enzyme that leaves blunt ends; the synthetic DNA molecule was then blunt-end-ligated to the cleaved plasmid. When the *lac* operon was induced, a protein was made consisting of the amino-terminal segment of β-galactosidase *coupled by methionine* to somatostatin, and terminated at the repeated stop codons of the synthetic DNA. This protein was purified and treated with cyanogen bromide, a reagent that cleaves proteins only at the

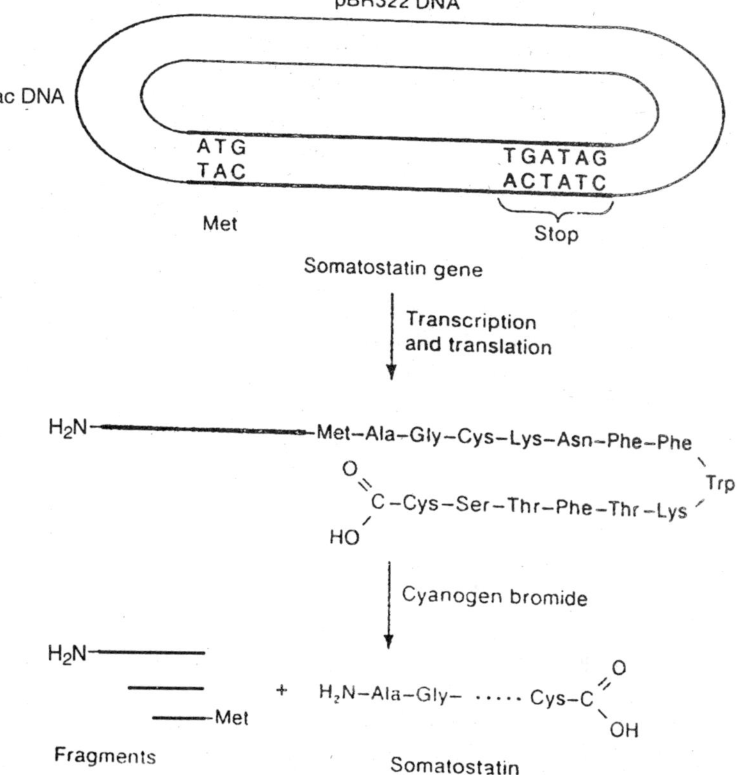

Fig. 11.16. Synthesis of somatostatin from a chemically synthesized gene joined to the plasmid pBR22 lac.

carboxyl side of methionine. In this way, the methionine linker remains attached to a β-galactosidase fragment and somatostatin is released. Use of methionine-coupling followed by cleavage with cyanogen bromide is a useful technique for separating any polypeptide from a bacterial protein to which it is fused, as long as the polypeptide itself does not contain methionine. Another more generally useful linker is $(Asp)_4$-Lys. In a sequence (Asp_4)-Lys-X, in which X is another amino acid, the enzyme enterokinase cleaves between lysine and X. Since such a sequence is not common, this linker is of more value than methionine.

Genetic Engineering with Animal Viruses

Retroviruses are RNA-containing animal viruses that have an unusual life cycle. These viruses, which contain the enzyme reverse transcriptase in their protein coats, use this enzyme to synthesize a double-stranded DNA copy of the viral RNA shortly after infection. As a normal part of the viral life cycle, this double-stranded DNA becomes inserted into the animal-cell chromosome, apparently at one of an enormous number of potential sites. Transcription occurs only after the DNA copy is inserted.

The infected host cell survives the infection, retaining the retroviral DNA in its chromosome. One of the best-understood retroviruses is *Rous sarcoma virus*, which causes tumors in chickens. Retrovirus DNA, either isolated from a cell or prepared synthetically, is useful vector with animal cells. Genetic engineering with retroviruses allows the possibility of altering the genotype of an animal cell. Since a wide variety of retroviruses are known, including two that can grow on human hosts, genetic defects may be correctable by these procedures in the future.

Many retrovirus species contain a gene that produces uncontrolled growth of a cell containing the retrovirus, thereby causing tumors in animals. If a retrovirus vector is to be used to change a genotype, the tumor-causing ability must be removed. This is possible by removal of the tumor-causing gene, which also provides the space needed for incorporation of foreign DNA. The recombinant DNA procedure employed with retroviruses consists of synthesis in the laboratory of double-stranded DNA from the viral RNA, through use of reverse transcriptase. The DNA is then cleaved with a restriction enzyme and, by the techniques already described, foreign DNA is inserted and selected.

Treatment of mutant cells in culture with the recombinant DNA, and application of a transformation procedure suitable for animal cells, yields cells in which recombinant retroviral DNA has been permanently inserted into an animal-cell chromosome. In this way, the genotype of the cell can be altered. Experiments have been done in which human cells deficient in the synthesis of purines have been obtained from patients with Lesch-Nyhan syndrome and grown in culture; these cells have been converted to normal cells by transformation with recombinant DNA. The exciting potential of this technique lies in the possibility of correcting genetic defects—for example, restoring the ability of a diabetic individual to make insulin or correcting immunological deficiencies. This technique has been termed *gene therapy*. However, it must be recognized that retroviruses are not well understood and are potentially dangerous. Gene therapy is not yet a practical technique, for major problems exist; for example, there is no reliable way to ensure that a gene is inserted in the appropriate target cell or target tissue. In addition, some means is needed to regulate the expression of the inserted genes. A major breakthrough in disease prevention has been the development of synthetic vaccines. Production of certain vaccines such as anti-hepatitis B has been difficult because of the extreme hazards of working with large quantities of the hepatitis B virus. The danger would be avoided if the viral antigen could be cloned and purified in *E. coli* or yeast, because the pure antigen could be given as a vaccine.

Several viral antigens have been cloned, but because of either thermal instability or poor antigenicity when pure, attempts to make vaccines in this way have generally been unsuccessful. However, the use of vaccinia virus (the anti-smallpox agent) as a carrier has been fruitful. The procedure makes use of the fact that viral antigens are on the surface of virus particles and that some of these antigens can be engineered into the coat of vaccinia. By 1985 vaccinia hybrids with surface antigens of hepatitis B, influenza virus, and vesicular stomatitis virus (which kills cattle, horses, and pigs) have been prepared and shown to be useful vaccines in animal tests. A surface antigens of *Plasmodium falciparum*, the parasite that causes malaria, has also been placed in the vaccinia coat; this may lead to an antimalaria vaccine.

Diagnosis of Hereditary Diseases

Restriction mapping has also been used in prenatal diagnosis—for example, in detecting sickle-cell anemia. This disease is caused by possession of an altered hemoglobin. The base change causing the mutation generates a cleavage site for a particular restriction enzyme. Cleavage of DNA from a sickle-cell patient produces a restriction map that differs from that of the DNA of a normal patient. The different restriction pattern can be detected by using Southern blotting and, as a probe, radioactive RNA prepared by transcribing cloned globin genes. The technique is so sensitive that the sickle-cell mutation can be detected in a small sample of cells obtained from the amniotic fluid surrounding a fetus, and detection is possible at a very early stage of development. Possibly in the future, gene therapy might be used to produce a normal child.

12

Molecular Scissor

Richard Roberts, a senior investigator at the Cold Spring Harbor Laboratory on Long Island, New York, is a collector. Not of stamps, or butterflies, or even beermats, but of enzymes. And the enzymes he collects are no ordinary enzymes; they are the machine tools of the genetic engineer, the precision cutters that allow the researcher to slice DNA cleanly at any known point. From time to time Roberts publishes his list in one of the learned journals; in 1978 he had about 160, and by 1980 the list had grown to 212. The list is now published each year in January, and Roberts fully expects the 1982 edition to top 300.

The published list grew out of a series of informal versions that Roberts circulated, which molecular biologists all over the world found enormously useful. They now help to keep the catalogue up to date by sending Roberts the information whenever they discover a new cutting enzyme. But what are these enzymes? Where did they spring from and how do they work?

Made in *E. coli*

Back in 1952, at the University of Illinois, Salvador Luria and his assistant Mary Human began to investigate a very interesting problem. The targets of their curiosity were bacteriophages, the so-called T phages, that infect the bacterium *Escherichia coli*.

Spread on to a nutritive agar jelly and kept warm, a thin suspension of *E. coli* cells will double every 30 minutes or so, quickly forming a smooth greyish yellow carpet over the surface of the jelly. But if the suspension is deliberately infected with bacteriophage, about one virus for every 10000 cells, and then grown as before, the bacterial lawn looks very different. Instead of being smooth and flawless, it is pockmarked by several small round holes. Each of these is called a plaque, and represents the site where a bacterial cell that was infected by virus came to rest. The DNA of the virus, injected through the virus's tail into the cell, subverts the bacterial machinery and forces it to obey the coded instructions of the viral genetic message, producing not bacterial components but viral components. These are assembled into new viruses and some time after infection, usually about an hour, the cell bursts and releases 200 or so new viruses. These infect neighbouring cells and repeat the process. By the end of 24 hours the myriad offspring of each successful virus have radiated out in a circle, destroying the bacteria around them and creating a clear plaque in the bacterial lawn.

There are many different sorts of phage, some labelled T2, T4, T6, T7 and so on, and many different strains of *E. coli*. Some strains of *E. coli* seemed to be protected against infection by certain of the phage types. Luria and Human discovered that growing T2 phage in one host often altered its properties so that it could no longer infect strains that it was usually quite at home in. But the altered phage was able to grow in dysentery bacteria (*Shigella dysenteriae*) and, most mysteriously, the offspring after growth in *Shigella* had regained their full infective power. The change that had come over the phage, whatever it was, was completely reversible, so it was certainly not a genetic mutation. It seemed to depend more on the host machinery that the previous generation had used to create the present viruses than on any property of the phage itself. Further experiments by Guiseppe Bertani in Luria's lab, and Jean Weigle at the California Institute of Technology, confirmed this strange discovery and added more details.

Bertani worked with a different phage, called lambda, which also attacks *E. coli*. Lambda grew beautifully on *E. coli* strain C, but the daughter phages coming out of strain C could not infect strain K. At least, not usually; one in 10,000 of the daughter phages from C did produce a plaque in a lawn of K cells, and everyone, rather than 1 in 10,000, of the viruses recovered from that plaque would now grow freely in strain K. They would also grow in strain C, but a single cycle of growth in C immediately changed the virus back so that it was again unable to grow properly in K.

At that stage, it looked as if there were two separate processes involved in the fight between bacteria and their virus parasites. Some

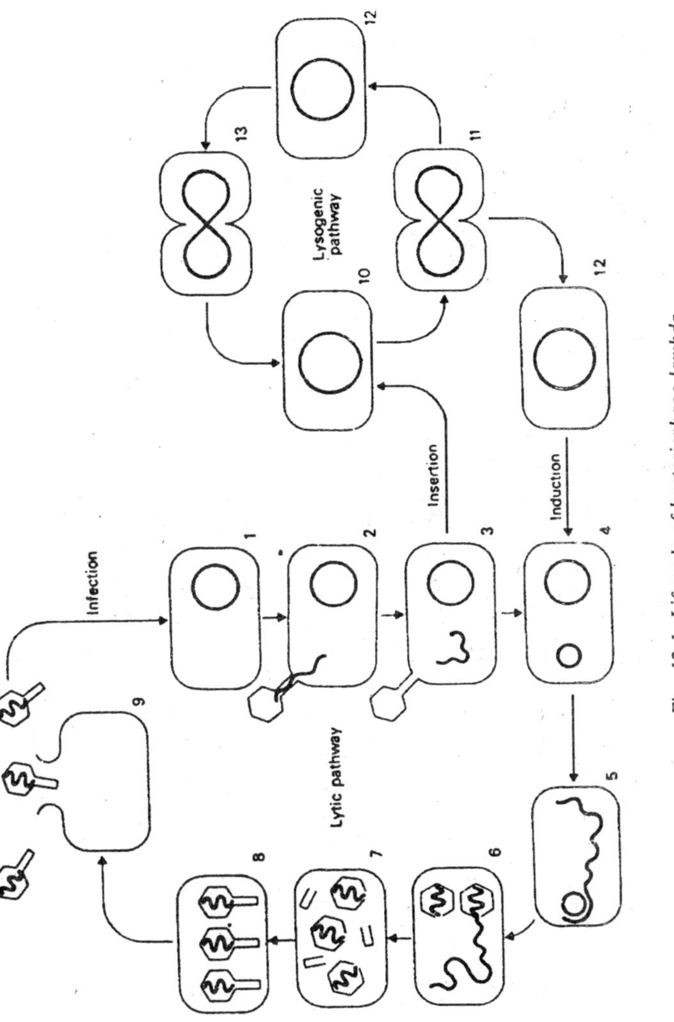

Fig. 12.1. Lifecycle of bacteriophage lambda.

strains of bacteria could successfully prevent virus attack. They were said to restrict the growth of certain types of phage. But if the phage could escape the restriction process, the survivors were somehow modified and able to infect the restrictive host freely. The infectiveness of a virus apparently depended on its most recent host. The systematic observations of Luria, of Bertani, and of Weigle introduced a measure of regularity to the strange phenomenon that came to be called 'host-controlled restriction'. These observations made sense, too, of other results that had been noticed, but not pursued, by earlier workers,

Public health workers need to identify disease-causing bacteria quickly and accurately to prevent epidemics. Different strains of bacteria are resistant to different antibiotics, require different vaccines and so on, and a valuable advantage can be gained by knowing the exact identity of the culprit before trying to treat the disease. One of the identification methods bacteriologists use is so-called 'phage typing'. They infect a lawn of the suspects strain with a set of 25 or more different sorts of bacteriophage and see which types of phage are able to grow and which are restricted. This alone will often tell the bacteriologist what strain he is dealing with. But sometimes things go wrong.

Ephraim S. Anderson and A. Felix worked at the Public Health Service's Central Enteric Reference Laboratory and Bureau in Colindale, north London. The major part of the laboratory's work was to identify the culprits behind various gastric upsets, often in cases of food poisoning. Prompted by Luria and Bertani's revelations of the mysteries of T2 and lambda, Anderson and Felix announced that in 1947, five years earlier, they had noted a similar thing with *Salmonella typhi*, one of the bacteria that causes food poisoning. Certain phages, called 'Vi-typing phages', were restricted by some strains of *S. typhi* and could grow on others. But if a phage was forced to grow on a restrictive host, it produced new phages that were no longer restricted on the original host. For Anderson and Felix this was a nuisance, because it meant that they had to keep their stocks of Vi-typing phages separate if they were to identify the *S. typhl* strains correctly; but in the light of the later work their observations made a great deal of sense. Anderson is, perhaps not unnaturally, a little bitter that he is seldom recognized as a pioneer in the field of restriction, but it is not that easy to make the connection between changing specificity – at that time thought to be a property of the virus – and restriction, a property of the host.

Some time after these events, Luria produced an explanation of the whole curious phenomenon of host-controlled restriction. He couched his account in literary terms. The genetic code, he said, was a universal language, each distinct triplet of bases along the DNA representing the same amino acid in all living things. So if one was looking at the code simply as a reader, the meaning would be the same regardless of who was doing the reading. Now suppose that, instead of regarding the code as a mere reader, one looks at it with the eye of a bibliophile. Little marks might, like the typeface or design of this book, reveal who constructed the code, who printed the book. Restricting bacteria were able to scrutinize all the DNA inside their walls, including that of the infecting phage. As a result they could tell that the phage DNA was foreign, and destroy it. To protect themselves, they would have to be able to recognize their own DNA, perhaps by modifying its structure in some way, and prevent it being destroyed. Occasionally a stretch of viral DNA would escape scrutiny and, while being replicated by the cell, would be modified and protected in the same way that the cell's own DNA is modified and protected. Instead of reading' made in *E. coli* (not strain K)', the DNA would now read 'made in *E. coli* strain K', and the vigilant restriction enzymes would be fooled. What the printer's marks might be, and how the foreign DNA was destroyed, remained mysteries.

Protection and Destruction

For roughly ten years from Luria's description of host-controlled restriction, the mechanics of restriction and modification remained unknown. Then, in 1962, Werner Arber, a Swiss biochemist at the University of Geneva, revealed what was going on. Arber had been inspired by a period spent with Jean Weigle in California, and he and Daisy Dussoix, a student working for her PhD, concentrated their efforts on a variant of phage lambda and two strains of *E. coli*, K12 and KP1. Their lambda phage would grow well in K12, but hardly at all in P1; less than 2 in 10,000 K12 phage particles survive P1, but those that do will breed equally well both in K12 and in KP1. Arber and Dussoix redid the experiment carefully, allowing the phage from KP1 to infect and burst just one K12 cell each, with no continued infection by the resulting daughter phage. The result was rather a surprise. Instead of just 2 particles in 10,000 being able to infect KP1, Arber and Dussoix got 1 in every 100 or so. Because the number of daughter viruses produced in a single cell before it burst was about 100, Arber and Dussoix argued that one progeny phage in each cell literally inherited the physical modification that had protected its parent.

This is complicated, so it is worth going through it again. Most phages grown on KP1 will not grow on K12, because they lack the made-in-K12 mark, but those that do grow will later grow in KP1 and K12. If you take a phage that grew first on KP1, then on K12, and allow it to infect a single KP1 cell, one in a hundred of the particles you get is able to infect K12. This may seem odd, but can be explained simply. The phage that grew in K12 picked up a made in K12' mark, and that mark is present when it infects KP1 again, and enables that particular phage to infect K12 in the future. Arber and Dussoix argued that one progeny phage in each cell inherited the physical modification that had protected the parent particle. Indeed, the infective 1 in 100 is the parent DNA, complete with its previously acquired 'made in K12' mark, repackaged into a new coat. On 8 December 1978, when his contribution to science was recognized by a Nobel prize, Arber explained their idea to an elegant Stockholm audience: 'We were convinced that this [modification] was transferred from the infecting parental phage particle. But was it a diffusible internal phage protein or was it perhaps carried on the parental DNA molecule?'

It turned out to be on the DNA, and to be a stable alteration to the molecule, not affected by the some times violent chemical procedures used to purify DNA. Finding out that the modification - which, as Luria had said, affected the substance of the DNA but not the spelling of the genetic message - consisted of tacking a small chemical group, a so-called methyl group, on to specific nucleotides in the DNA, took a little longer - until the early 1970s, in fact. When Arber grew *E. coli* on a medium deficient in methionine, the amino acid that cells use as a source of methyl groups, the bacteria lost their ability to modify DNA. Indeed, other experiments suggested that cells would not survive long without methionine because their own DNA was unprotected and so, in a sort of inevitable suicide, the cell's restriction enzymes turned on the very DNA that had created them.

While Arber was working out that modification involved changes to the DNA, Dussoix and another colleague, Grete Kelenberg, tried to find out what restriction really was. They grew phages on strain

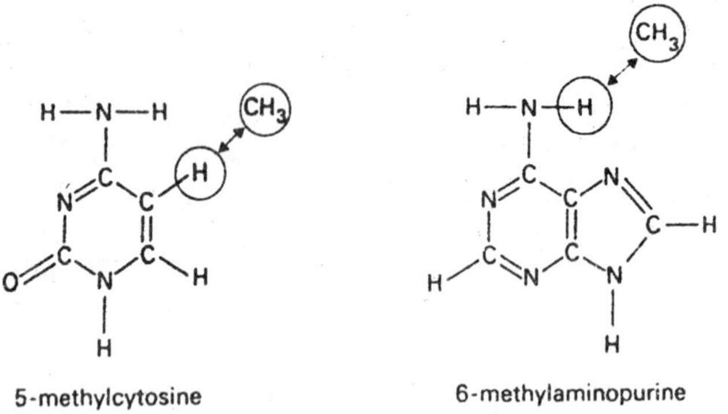

5-methylcytosine 6-methylaminopurine

Fig. 12.2. Modified bases. The substitution of a methyl (CH_3) group for one of the hydrogen atoms in a base marks that base out and protects the DNA from restriction enzymes.

K12 bacteria which had in turn grown up in a medium that contained ^{32}P, a radioactive isotope of the element phosphorus. As a result, the phage DNA contained a radioactive label and could be traced. Dussoix and Kellenberg infected various strains of *E. coli*, some restrictive and others not, with the labelled phage. After an interval of 3, 7 or 15 minutes they stopped all chemical reactions in the cells and looked to see what happened to the radioactive label. On strain K12, which of course did not restrict the growth of phage that had been grown on K12, over 90 per cent of the hot phosphorus was in the form of large insoluble molecules inside the bacteria. On P1, which did restrict phage growth, more than half the phosphorus was floating freely in solution and obviously no longer part of a DNA molecule. Even after as little as three minutes, unmodified DNA injected into restrictive cells was completely ineffectual, having been broken down into tiny bits.

The Enzyme Isolated

After Chicago and California, where he had pioneered density gradient methods for separating large molecules, and had been part of the team that isolated messenger RNA, Matthew Meselson found himself with his own lab at Harvard. In 1966 Robert Yuan, a Fellow of the Damon Runyon Memorial Foundation, arrived as a postdoctoral fellow. Meselson offered Yuan a choice of two projects. Either he could investigate the enzymes involved in the recombination of phage T4, or he could search for the host restriction enzymes. Undaunted by previous failures, and encouraged by Meselson, Yuan decided to make a frontal attack on restriction enzymes. They chose to purify the enzymes responsible for restriction in an *E. coli* of strain K, its code number in the American Type Culture Collection being 1100. A mutant derivative, number 1100.293, lacks the genes that make the restriction enzyme and hence does not have the ability to destroy foreign DNA.

To pursue the purification, they needed a way to measure the restriction power of any extract, and Meselson adapted the density gradient method. DNA, being a relatively heavy molecule, sinks slowly if it is centrifuged in a solution of sugar, but the compact twisted circles of lambda phage DNA sink much more quickly than DNA that has been opened out. Lambda DNA, treated with a crude extract of 1100.293 (that is, unrestricted), sank 2 cm in an hour and 48 minutes; by contrast, it sank only 1½ cm in the same time if treated with an extract of the restricting strain. Clearly, the restricting strain did have the power to open up the DNA and make it sink more slowly. The assay method was a bit tedious: labelled DNA had to be tracked through the density gradient by counting the particles emitted by each successive millimeter of sucrose solution; but it worked, and revealed when DNA had been broken up by enzymes and when it had not.

Purification began, as it almost always does, with a technique called 'dialysis'. The rag-bag mixture of proteins and everything else from the cells is put into a sack made of a special material and submerged in a mixture of solvents. The small impurities pass through the membrane and the dialysate, as it is called, that is left behind is a purer solution of the large proteins. (Dialysis, essentially, is like washing; the technique is in everyday use to clean kidney patients' blood of harmful impurities.) Unfortunately for Meselson and Yuan, the dialysate of *E. coli* K seemed to have no restriction activity. DNA was untouched by it. This stumped them for a while and caused a frustrating delay, until they discovered that two other chemicals, so-called 'co-factors', were needed for the restriction enzyme to work. One of these was ATP, the molecule that stores and transports energy in the cell. The other was S-adenosylmethionine, SAM for short. As SAM provides methyl groups, and is needed for good restriction in living cells, it was no great surprise that the cell extract also needed it. ATP and SAM provided the crude cell extract with quite a powerful destructive effect on lambda DNA.

With the requirements of the reaction all worked out, Meselson and Yuan proceeded to the business of purification. They grew *E. coli* in vast amounts, until the stock broth was a soup containing 600 million cells in every millilitre. The stock culture was spun down 'in a centrifuge, the spent broth thrown away, and the cells stored at -20°C. From the frozen mass, they took 120 grams of cells and whirled them in a blender with tiny glass beads. The beads, like a miniature crushing mill, broke open the cells and released the cytoplasm. Two hours at 35,000 r.p.m. separated the empty cell walls from the cytoplasm, which tests showed was able to degrade DNA.

Cytoplasm contains a great deal more than restriction enzyme; apart from all the other enzymes, there are assorted other chemicals that are not wanted. The 300 millilitres of crude extract now had to be cleaned up. The first stage was to separate the proteins from the other chemicals. Ammonium sulphate is gentler than phenol and does this nicely, making the proteins precipitate out of solution. Meselson and Yuan collected and precipitated protein and dissolved it up in a special mixture, which they dialysed to get rid of small impurities. Then they purified the dialysate by column chromatography.

The principle of column chromatography is simple. A mixture in this case of proteins, is poured on to the top of a column of some absorbent material. Meselson and Yuan used chemically treated cellulose. The proteins tend to stick to the cellulose, but they can be washed off again by the buffer solution they are dissolved in. Each protein has a characteristic stickiness that determines the speed at which it will travel through the column. Stickier proteins get left behind, while not so sticky ones are the first to drip out of the end of the column.

Meselson and Yuan poured the dialysate, 100 ml of it, on to a column, 6 cm across and 11 cm deep, of specially prepared cellulose. The proteins sorted themselves out, and the two researchers washed the column with a litre of buffer solution, taking care to collect separate 50 ml samples, or fractions, as they dripped through. The restriction activity was sure to have collected in one or more of those 200 samples, but to find out which they had to go through the tiresome business of incubating some of each sample with DNA, putting the DNA on to a sucrose gradient, and seeing whether the DNA had been attacked. When it was all done the results were clear. All the activity was concentrated in just two of the fractions. The rest could be thrown away.

The two fractions were mixed together, dialysed again, and put through the chromatographic separation once more. This time the column was taller and thinner, 1 cm by 15 cm, and made of slightly different cellulose, all of which combined to enhance the separation yet further. Washing the samples out now used only 300 ml, and each fraction was only 9 ml; but again, each fraction had to be tested to see whether it had the power to restrict DNA. The restriction activity was concentrated into just three adjacent 9 ml fractions. The fractions were pooled, dialysed, and concentrated down to 7.5 ml by removing water. Chromatography now was too insensitive to separate the constituents of the mixture, because if there were several proteins they were essentially all equally sticky. But they might be of different density, so the final step was a purely physical separation using a density gradient.

The 7.5 ml was divided into three portions and each was placed on top of a test-tube containing a density gradient of glycerol. The tubes were spun at 25000 r.p.m. for 20 hours and gingerly removed from the centrifuge. A pin-prick in the bottom of each tube allowed the glycerol and layered proteins to drip out each millimeter of liquid in the tube collected as a separate fraction and tested. Three fractions had the power to break up DNA; they were over 5000 times more active than the crude cell extract.

Those days, Yuan recalled, formed 'one of the most exciting periods in my life. 'The long hours, made the finding of open eating places an added complication.' At the end of it, with 'great joy', he and Meselson published their work in *Nature*. From the 8 grams of protein in the original extract, Meselson and Yuan had obtained something like, half a thousandth of a gram, but that tiny amount contained practically all the restriction activity of the 120 grams of *E. coli* cells. With pure enzyme, they could look in detail at how it worked.

Destroyer and Protector

Meselson and Yuan's enzyme was technically an endonuclease. That is, it acted within a double helix of DNA to break it, rather than attacking the ends of the helix. When they attacked labelled DNA with the enzyme, very little of the radioactive phosphorus found its way into the solution. Most of it stayed bound up in the DNA, which proved that the enzyme did not just break up the DNA molecule by chomping away at one end. Also, the enzyme worked perfectly well on the closed circles of DNA that comprise the genetic message of lambda phage. So it attacked double-stranded, whole DNA.

The product of the pure enzyme was also double-stranded DNA, though obviously in shorter lengths. Incubating a mixture of DNA and enzyme extensively and then adding more enzyme did little to alter the pattern of settling of the DNA in the sucrose gradients, a pattern that resembled the one that would be obtained if the entire lambda DNA were divided into four quarters. Another' trick proved that the degraded DNA was double-stranded. Normal double helices settle at the same speed in strong and weak salt solutions, whereas single-stranded DNA curls up in strong salt solution and settles three times more quickly than in a weak solution. Meselson and Yuan found that the enzyme-treated DNA settled at the same speed regardless of salt concentration; it was double-stranded. Furthermore, there are no single-strand breaks, or nicks, in the treated DNA. Separating the two chains does not alter the settling pattern, as it would if some of the strands were nicked into shorter pieces.

It looked as if the restriction enzyme from *E. coli* K worked in two stages. It made a break in one strand and then, some 30 seconds later, the second strand broke. Meselson and Yuan couldn't tell whether the same enzyme molecule made both breaks, nor were they sure whether the enzyme was attacking a specific sequence along the DNA.

The pattern of fragments that they got suggested that the enzyme might recognize just a few sites that it then attacked, but they couldn't be certain that this was so.

Perhaps the most interesting result was that a modification on just one of the two strands protected the molecule from the enzyme. DNA molecules from phages grown in K were modified and protected against the enzyme of K. Those grown in C were not. Meselson and Yuan separated the DNA strands in samples from the two types of phage and then mixed the single strands. Some of the DNA from K-grown phage came together with strands from C-grown phage to form hybrid double helices, known as heteroduplexes, on which one strand, the K strand, was protected and the other not. The enzyme did not attack these hybrid strains, not even to make a nick in just the unmodified strand. This finding was extremely important, as Meselson and Yuan realized:

> The resistance of heteroduplexes may serve to protect newly replicated bacterial DNA from attack by the cells' own restriction enzyme, allowing time for modification of the newly synthesized chain. Indeed, if restriction and modification are accomplished by the same enzyme, the choice between the two reactions may normally be governed by' whether the substrate is an unmodified homoduplex or a heteroduplex, respectively.

Perverse Complexity

The achievements of Matthew Meselson and Robert Yuan in isolating and purifying the restriction enzyme of *E. coli* and working out how it operated were enormous. But it was only in 1979 that we gained a understanding of how it works. The reason for the delay was, as the writers of one textbook put it, that the enzyme Meselson and Yuan chose to work with is 'perverse in the complexity of its behaviour'. The final story is astounding.

The enzyme is large, and is made of three parts. One recognizes a specific sequence of bases on the DNA. The second cuts the DNA if, and only if, the recognition site contains no protective methyl groups. The third part adds methyl groups to one strand if the other strand already has one. The recognition site is 13 nucleotides long:

5′ —A—A—C—N—N—N—N—N—N—G—T—G—C— 3′
3′ —T—T—G—N—N—N—N—N—N—C—A—C—G— 5′

The middle section of six N's indicates that the exact identity of those nucleotides is irrelevant, but the flanking sequences of three and four are vital. How the enzyme recognizes the site is a mystery, although it looks as if the recognition part of the molecule fits inside the groove of the double helix and reads its way along. If, when it comes to a recognition site, it finds a methyl group on one strand (the exact location of the protective methyl is still unknown), it acts as a methylase and adds protection to the other strand. If it finds both strands methylated, it stops reading and falls off the helix. If it finds no methyl groups it destroys the helix.

The exact details of destruction still aren't known for the enzyme from *E. coli* K, but a very similar enzyme from strain B is better understood, John Rosamond worked it out with some colleagues while he was a visiting researcher at Berkeley, using direct observation with an electron microscope, and a series of cleverly contrived stretches of DNA (manufacture of which, ironically, had to await developments in the technology of restriction enzymes). The DNA in question had target sites for the enzyme at different distances from the end of the molecule. Only if the site was more than 1000 bases from the end of the DNA did the enzyme cut it. And it never cut the DNA at the recognition site; the damage was always at least 1000 bases from the site.

The enzyme becomes bound to the recognition site and undergoes some sort of transformation, perhaps losing one of its subunits. Then, while still attached at that site, it begins to pull the DNA through itself. It always pulls in DNA from the same side of the sequence, the side with three recognition bases, holding on to the four-base site and effectively, travelling along the DNA. Looking at the complex of DNA with enzyme in the electron microscope, Rosamond could see loops at the end of the molecules. These loops show the enzyme frozen in the act of walking along the DNA. When it has pulled in a certain amount of DNA, somewhere between 1000 and 5000 nucleotides, it stops. Nobody knows why it stops where it does; certainly, the site of the break is not fixed either with respect to the recognition site or with regard to the sequences of bases there. Perhaps it is as simple as a kink in the DNA. In any case, regardless of the reason why the enzyme stops tracking along the molecule, once it has stopped it makes a nick in one of the strands. Other enzymes come along and attack the exposed DNA in the nick, dissolving away a stretch of about 75 nucleotides before the second strand is broken. The result is two fragments of DNA that can be swiftly degraded by yet more enzymes. And long after the DNA has been severed the restriction enzyme remains bound to the recognition site, where it breaks down large amounts of ATP. Some ATP is used to power the hauling in of the DNA through the enzyme, but why the cell continues to waste energy molecules with such profligacy is not clear; it might be a man-made artefact caused by lack of some vital regulatory molecule in the test-tube cell substitute, or it may represent a specific extra function for the restriction enzyme that is at present unknown.

There are undoubtedly further twists to be unravelled in the byzantine knot of restriction enzyme activity. And all that complexity, while undoubtedly fascinating, hardly seems worth all the effort. What use is a cutting tool that, while it recognizes a specific site, nevertheless makes itself felt at some random point 5000 bases away from that site? The answer, quite simply, is that, aside from its intrinsic interest, the restriction enzyme that Meselson and Yuan purified from *E. coli* K prompted a search for other restriction enzymes, and the vast majority of these turned out to be much simpler. They recognize a sequence and, without any need for ATP or SAM, break the DNA right there and then. No tracking, no randomness, just a nice clean slice. And tools like that are very handy indeed.

A Whole New Class

Sharing the Nobel platform with Werner Arber in December 1978 were Hamilton Smith and Daniel Nathans. Smith and Nathans are both professors of microbiology at the prestigious Johns Hopkins University School of Medicine in Baltimore, Maryland. Where Arber had pioneered the detailed investigation of the enzymes that lay behind host-controlled restriction, Smith was honoured for discovering a whole new class of restriction enzymes, and Nathans for putting those enzymes to work in the service of fundamental molecular biology.

It began, as success stories often do in science, with a lucky chance. Smith, at the time a recently arrived member of the Johns Hopkins faculty, was working with bacteria called *Haemophilus influenzae* strain Rd. This had the useful property of being very changeable, and Smith began to investigate some of the transformations that the bacteria underwent as they swapped bits of DNA with other bacteria. With a young graduate student, Kent Wilcox, Smith used a viscometer to measure the extent to which cell extracts broke up DNA. Long strands of DNA from a very viscous solution, while smaller ones are less viscous; a decrease in viscosity mirrors the break-up of DNA. As another part of the study, Smith allowed the *H. influenzae* cells to take up radioactively labelled DNA from various sources and then tried to recover this donor DNA to see what had happened to it.

One day, Smith and Wilcox offered the bacteria labelled DNA from a phage called P22, a virus that Smith had long been familiar with. To their surprise, when they came to look for the labelled DNA in the cells, they couldn't find any.

When he received his Nobel prize Smith continued the story for his audience in Stockholm, trying hard to conceal the emotion involved:

With Meselson's recent report in our minds, we immediately suspected that the phage DNA might be undergoing restriction, and our experience with viscometry told us that this would be a good assay for such an activity. The following day, two viscometers were set up, one containing P22 DNA and the other, *Haemophilus* DNA. Cell extract was added to each and we quickly began taking measurements. As the experiment progressed, we became increasingly excited as the viscosity of the *Haemophilus* DNA held steady while that of the P22 DNA fell. We were confident that we had discovered a new and highly active restriction enzyme.

Their confidence, needless to say, was not misplaced, but it took a considerable amount of effort to find out what they had and how it worked.

Smith and Wilcox went through more or less the same stages that Meselson and Yuan had gone through; there is really no alternative, and the purification of almost any protein from a cell would be bound to involve many of the same procedures. The starting point was a culture of *H. influenzae*, growing in a 12 litre vat of sustaining brain and heart broth. Instead of glass beads, Smith used ultrasonic vibrations to break open the cells and release the cytoplasm. Put in the viscometer with P22 DNA, just ten-millionths of a litre of the crude cell extract was enough to start breaking up the DNA and reducing the viscosity. The crude cell extract was washed, separated chromatographically, washed again, and separated again. At the end, having left out the purification by density gradient that Meselson and Yuan had used, Smith and Wilcox had an enzyme that was 200 times more active than the raw cytoplasm.

This restriction enzyme needed neither ATP nor SAM, though it would not work unless there were magnesium ions present. It was smaller than the *E. coli* K enzyme, with a molecular weight of only about 70,000. It attacked foreign DNA with alacrity, reducing the genome of T7 from a single length of molecular weight 26,400,000 to about 40 bits with an average molecular weight of 1,450,000. The bits were all double-stranded, and no nucleotides were removed from the chain. The enzyme appeared to chop through the DNA in one go producing nicks in the two strands at almost the same time and place. They christened the enzyme 'endonuclease R'; endonuclease because, like Meselson and Yuan's enzyme, it attacked nucleic acid molecules from within (that is, it could break the middle of a chain – exonucleases attack at the tips of the molecules), and R for restriction.

Meselson and Yuan had guessed that their enzyme recognized a specific sequence of bases. For similar reasons, namely the relatively small number of breaks produced, Smith and Wilcox felt the same way about their enzyme. But they went further and estimated the length of the recognition site. T7 DNA is about 40,000 bases long, and the *Haemophilus* enzyme broke this into 40 pieces about 1000 bases long. On a stretch of randomly organized DNA a particular unique sequence of, say, four bases would occur roughly every 256 bases; clearly, four bases is too small a site. To occur every 1000 or so bases, a site would need to be five or six bases long. At the very end of their paper announcing the new restriction enzyme, Smith and Wilcox derived this estimate and then had the satisfaction of stating: 'In the accompanying paper ... the base sequence recognized by endonuclease R is completely identified and provides confirmation of this estimate.

Recognition

At the same time as he had been working on the properties of the *H. influenzae* restriction enzyme, Smith was also tackling the question of its recognition site. Kent Wilcox was in on the early stages of this project too, but could not continue because the army had sent him his draft notice of induction. His prime preoccupation therefore become to complete the formal requirements of his master's degree. Smith continued alone for a time, and then was joined by Thomas Kelly Jr.

All the evidence pointed to a specific recognition site. Of the 40000 sugar-phosphate links in each chain of the viral DNA, the enzyme broke only 40. There had to be some sort of guidance mechanism to get the enzyme to cleave just there. The strategy Smith and Kelly adopted sounds simple. They would look at the ends of the fragments and identify the nucleotides there. If the site was specific, they would find a specific pattern of nucleotides ending each and every fragment. Putting it into practice required a battery of other enzyme systems, two sorts of radioactive label and a great deal of ingenuity.

The starting point was purified endonuclease R and T7 DNA labelled with ^{33}P. This is a radioactive form of phosphorus that is less energetic than the usual ^{32}P; it was used as a general label to tell them where the DNA was. Smith and Kelly attacked the single nucleotide at the 5' end of the fragments first. An enzyme called alkaline phosphatase removes the phosphate group from the end of a fragment. This done, the fragments were incubated with ATP labelled with energetic ^{32}P and another enzyme, polynucleotide kinase. This enzyme transfers the labelled phosphorus from the ATP to the nucleoside. After an hour at blood heat the doubly labelled DNA – ^{33}P throughout and ^{32}P at the 5' end – was extracted, purified and analysed.

Snake venom is the source of yet another enzyme, this one called snake venom phosphodiesterase. This has the useful property of breaking DNA into its component nucleotides, leaving the phosphate group attached to the 5′ carbon of the sugar. When the venom enzyme had done its work Smith and Kelly had a mixture of four different nucleotides, all radioactively labelled but some – the nucleotides that were at the 5′ end of the restriction fragment – labelled with ^{32}P while all the others carried only ^{33}P. The problem now was first to separate the four nucleotides, and then to identify the ones that had the hotter ^{32}P label. Separation was routine, using electrophoresis. Similar to chromatography, electrophoresis is a process in which an electric current frog-marches the molecules through a thin layer of cellulose. Molecules that are small, or highly charged, move faster than those that are large, or not so highly charged. Once separated, the nucleotides had to be identified. This is where the differential labelling came in. ^{33}P is not very energetic, and the particles it gives off as it decays are easily blocked ^{12}P, on the other hand, gives off more energetic particles. Smith laid a sheet of Kodak X-ray film on either side of the cellulose gel that had been used for electrophoresis. As the phosphorus decayed it gave off particles, which collided with the silver grains in the film and left a permanent record of their position. The radioactive phosphorus essentially takes a picture of itself; hence the name autoradiography. One side of the gel was in direct contact with the film. Strong and weak particles could get to the silver, so ^{33}P and ^{32}P each made their mark. On the other side of the gel was the gel's thin supporting plastic. Flimsy though this was, it was enough to stop the weaker decay products of ^{33}P. Only the hotter label of the terminal nucleotide would show on this film.

Conceptually simple, this intellectual breakthrough was absolutely fundamental to the work that was to follow. But after the very first run, when they looked at the films and compared the exposed positions with the pattern they got with known nucleotides. Smith must have been a bit disappointed. The hot label was present in two nucleotides 63 per cent attached to the adenine and the remaining 37 percent to guanine. So the cleavage site was 'not absolutely specified'. All was not lost, however, for adenine and guanine are both purines; the enzyme made a break on the 5′ side of a purine, but didn't distinguish between adenine and guanine.

What was the penultimate nucleotide on the fragment? As before, the phosphorus on the end was replaced with a hot phosphorus, but instead of using snake venom, an enzyme called exonuclease I was brought in. This nibbles away at DNA fragments from the 3′ end, but stops when there are just two nucleotides left. Again, the dinucleotides were separated and auto-radiographed. The label was again in two places. One, with 62 per cent of the radioactivity, could have been either AA, or CG, or GC. The other was either GA or AG. Snake venom separated the first pair into single nucleotides; all the hot label was on adenine, so the pair was definately AA. The second pair's ^{32}P was also on the adenine, so this one was GA. The site was specific. The 5′ base could be either A or G, but the next base along was always an adenine.

The final three bases on the 5′ end of the fragment turned out to be AAC and GAC, so the specificity was holding up. What about the other side of the break, the 3′ end of the fragments? Here another bit of thought enabled Smith and Kelly to take a shortcut. There were, logically, only two ways that the enzyme could work. It could either break the DNA duplex evenly, producing fragments with flush ends, or it could break unevenly, so that the fragments had one longer strand at the end. Only if the break was even, and the ends flush would the sequence of bases at the 3, end of the fragment be exactly complementary to the sequence at the 5′ end. Yet another enzyme – micrococcal nuclease was brought in. This enzyme breaks whole DNA into single and double nucleotides, those from the 3′ end being double nucleotides joined by a single phosphate. Smith and Kelly separated these out from all the other nucleotides by chromatography and found that they were of two different sorts. Breaking them up and looking at the individual nucleotides revealed that one was TC and the other TT. This was exactly what they would predict if the break is even. They had almost all the details of the recognition site: it was

5′ —G—T—Py—Pu—A—C— 3′
3′ —C—A—Pu—Py—T—G— 5′

with the break coming between the two unspecified bases in the middle. (Notice that the sequence is a palindrome. It reads the same on the two strands, just as the sentence 'able was I ere I saw Elba' reads the same forwards or backwards. The vast majority of subsequently identified recognition sites turned out likewise to be palindromic: the consequences of this are extremely important.) Such a sequence of six bases would occur in a random genome once every 1024 base-pairs. The fragments T7 DNA produced by endonuclease R were about 1000 base-pairs long, while those of P22 were 1300 base-pairs long. As Kelly and Smith noted:

These facts suggest that the sequence [above] contains sufficient information to account fully for the observed degree of specificity of endonuclease R. However, since strictly speaking neither the T7 nor the P22 DNA molecule represents a truly random collection of nucleotides, the possibility remains that the recognition region for endonuclease R might be larger·

They had to look at the fourth nucleotide from the end; if that too was specific, the recognition region was longer than six base-pairs. Fortunately, it was not.

The recognition site of a restriction enzyme had been fully identified, and the pace of progress was speeding up. Thirty years from the first phage-typing to Luria's experiments; ten years from Luria to Arber; eight years from Arber to Meselson and Yuan; and two years from them to Smith. Host-controlled restriction, which Hamilton Smith described as 'an apparently insignificant bacteriological phenomenon', had yielded an enzyme that, he saw, had 'unexpectedly far-reaching implications'.

Things Get Easier

Almost as soon as Smith had discovered the first specific restriction enzyme, his colleague Daniel Nathans was putting it to work in the service of molecular biology. Nathan's target was a little virus that infects the cells of monkeys, hence its name simian virus 40, or SV40. SV40 can, when conditions are right, transform the cells it has infected and make them grow like cancerous tumours, so it has not unnaturally been the subject of intense interest by many groups of researchers. But before examining the developments, in SV40 and elsewhere, that followed Smith's work with the *Haemophilus influenzae* enzyme, it is worth looking at the way in which interest in restriction enzymes took off.

Meselson and Yuan had an enormously hard job demonstrating that the various fractions they collected were, or were not, possessed of restricting activity. They had to use modified DNA labelled with one type of radioactive isotope and unmodified DNA with a different label. Then, after incubating a mixture of the two with the putative enzyme, they had to spin the products down a sucrose density gradient. Dividing the gradient into portions and counting the different types of radioactivity in each took yet more time, and at the end the results had to be examined carefully to see which fractions were able to destroy unmodified DNA. It worked, but it took time.

Smith and Wilcox went one better. No longer were they dependent on tedious, density-gradient separations and radioactivity assay (though they still used those for confirmation); they had the viscometer, which gave a purely physical indication of whether the DNA was being broken or not. In the right hands, this was a much speedier process for testing the various fractions, and enabled Smith and his helpers to progress rather rapidly. Still, the techniques were not as fast as they were to become, and it was Nathans and his SV40 who pushed the speed up another notch.

Nathan's intention was to divide the DNA of SV40 up into manageable fragments, the better to understand how the entire virus worked. He quickly established that Smith's restriction 'enzyme did indeed break open SV40's circle of DNA, and that it created several fragments, The next goal was to resolve the mixture of fragments into its component parts, and for this Nathans enlisted the help of Kathleen Danna, a graduate student in his laboratory. The technique they used was electrophoresis, but rather than using cellulose as a supporting medium for the mixture of fragments they used a gel containing polyacrylamide. This is a long thread like molecule that links to form a mesh permeated by small pores. The amount of polyacrylamide sets the average size of the pores, which in turns restrict the passage of molecules through the gel. Small molecules always travel faster, but by adjusting the amount of polyacrylamide in the gel the biochemist can fine -tune his molecular sieve to sort out a particular size most efficiently: Nathans and Danna put the sample of radioactive SV40 DNA, digested by the restriction enzyme, at one end of a 13 cm strip of polyacrylamide gel. The DNA, of course, is completely invisible, so to keep track of it they added a dye, bromophenol blue, that moved little faster than the smallest chain of DNA. Sixty volts from one end of the gel to the other provided the motive power to separate the fragments of DNA, smaller fragments racing ahead of larger ones, and at the end of eight hours the blue band of dye had travelled 13 cm and reached the end of the gel. The problem now was to make the invisible visible, and see what had become of the DNA.

Danna froze the 13 cm strip of gel to make it easier to handle and then cut it into about a hundred bands, each just 1.2 mm wide. Each band sat for a while in solvent, and then the amount of radioactivity that had leaked out into the solvent was counted with a superior sort of geiger counter. The results showed clearly that the fragments of nucleic acid had indeed separated out. The column of numbers counts of radioactivity from 100 or more samples, produced a clear pattern of peaks and troughs; nine peaks and ten troughs. Two of the peaks contained more than their fair share of DNA, an anomaly that autoradiographs of the gels explained by showing two bands close together at each of the two positions. The restriction enzyme from *H. influenzae* chopped the DNA from SV40 into 11 chunks, and a whole series of different measurements proved that the amount of DNA in the 11 fragments added up to the total amount of DNA in the whole virus. 'We conclude,' wrote Danna and Nathans, 'that every molecule of SV40 DNA yields one of each of these fragments. That was why each of the fragments formed a band in the polyacrylamide gel. They were always the same set of fragments, and they lined up in the gel according to size. It was just the result Nathans needed.

It was also ideal for identifying new restriction enzymes. If some portion of purified cell extract gave evidence of breaking DNA, the products of the digestion could be layered on a gel and separated by electrophoresis. If the enzyme was truly a restriction enzyme, breaking the DNA at a set number of specific sites, then it would produce the same set of fragments from every molecule of DNA, and the fragments would line up to form the tell-tale bands. One had to use radioactive DNA, it's true, and follow up the electrophoresis by slicing up the gel and counting the radioactivity, or else by allowing the fragments to photograph themselves; but Danna and Nathans had made it possible to look for restriction activity even more efficiently than had Smith and Wilcox.

Another advance came a couple of years later, in 1973, from Cold spring Harbor Laboratory. Phillip Sharp, Bill Sugden and Joe Sambrook streamlined the search for restriction enzymes still further. Their breakthrough was in two parts. First, they changed the electrophoresis gel, from polyacrylamide to agarose. DNA fragments are large molecules, and a polyacrylamide gel with holes large enough for the DNA to get through contains so little polyacrylamide that it forms an impossibly mushy gel. Agarose, a long chain polymer rather like starch, can provide additional support for a weak polyacrylamide gel and is better than polyacrylamide because agarose is not as sensitive to the concentration of electrically charged particles in the solution. When a protein mixture of enzymes has been separated chromatographically the different components are washed out of the column with different concentrations of solvent. Previously, each fraction had to be adjusted to the same concentration before being tested with DNA and electrophoresed. With agarose, the fractions could be taken straight from the column, incubated with DNA, and put through electrophoresis.

The second improvement was perhaps even more important, for, it got rid of the lengthy procedures needed to make the DNA in the gel visible. Slicing and counting, staining with dyes (which needed to be removed again before anything else could be done with the DNA), and autoradiographs were all time-consuming and tedious. Sharp added a simple chemical, ethidium bromide, to the gel and the mixed fragments of DNA. Ethidium bromide inserts itself between the bases of the DNA and stretches the helix out slightly, but that is neither here nor there. Much more important, the DNA, makes the ethidium bromide glow in ultraviolet light. No staining, no radioactivity, no lengthy wait for autoradiographs; you simply, take the agarose gel into a darkroom, switch on an ultraviolet light, and there it is. Bands containing as little as 50 billionths of a gram (0.05 microgram) of DNA light up yellow-orange; to the eager researcher the glowing bands can seem as bright as neon signs and just as conspicuous. Sharp's technique made searching for restriction enzymes a dream Or, in the dry prose of a scientific paper: "the lengthy staining and destaining or autoradiographic procedures which are an integral part of most of the techniques that are used for electrophoresis of DNA are eliminated ... In our hands the technique has proved to be a very useful and flexible tool for assaying restriction enzymes."

Sharp's method made everything much easier; separating and seeing DNA was now relatively simple, and the community adopted the method with alacrity. 'Agarose gel electrophoresis did not originate with our *Biochemistry* paper, 'Sharp has told me, 'but was certainly popularized by it'. The DNA detection was another matter. Sharp had spent a year with a postdoctoral fellowship at the California Institute of

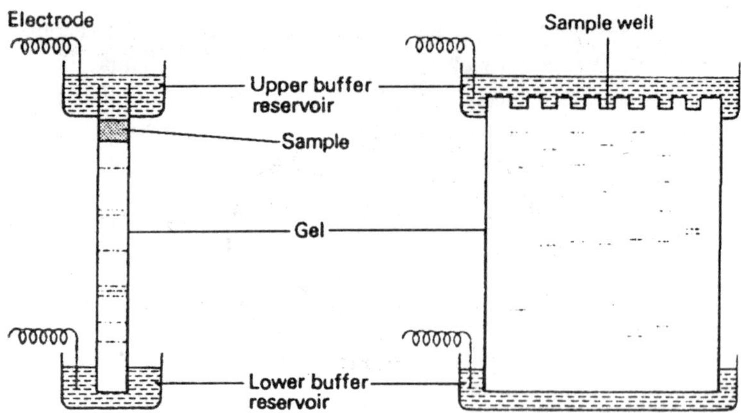

Fig. 12.3. Electrophoresis apparatus. The sample mixture is carried through the gel by an electric current, which separates out the fragments according to size.

Technology, where he had seen how the dye ethidium bromide fluoresced much more brightly when bound to DNA. He remembered this at Cold Spring Harbor when he wanted to visualize DNA fragments.

I decided to try to stain gels with ethidium bromide as a means of detecting the bands formed by different length fragments. The minimum concentration of ethidium bromide necessary for saturation of DNA was calculated and within 3 hours I had successfully detected DNA bands in gels with ethidium bromide.

Another conceptual leap came from Camil Fuchs, a statistician at Wisconsin university. Fuchs made it vastly easier to discover the recognition site of a new restriction enzyme. Working from the starting point that the enzymes recognize a palindromic site, Fuchs wrote a computer program to detect 4-, 5- and 6-base palindromes, and set the program to work on the complete sequences of two viruses, SV40 and ΦX174, both of which were known in their entirety. The computer told Fuchs how many fragments would be generated, what length they would be and so on for every palindrome in the two genomes. To discover the enzyme's most likely recognition sequence, the researcher had only to use the new restriction enzyme on SV40 and ΦX174, measure the fragments so easily separated and visualized, and simply read the tables that Fuchs published in *Gene*. It was a simple matter then to confirm that sequence by more pedestrian methods.

The contrast between then and now, between sucrose gradient separation and fluorescent bands, is hard to imagine. Robert Yuan remembers that in those early days, such assays served their function well, but they were arduous, time-consuming, and extremely boring. The intellectual elegance of some of the experiments was tempered by the laboriousness of many of the methods'. Now, nobody gives a second thought to the ease with which results can be obtained. There's no substitute for elegance, which still needs to be there, but the tedium is not quite so over- whelming.

The Naming of Parts

When Sharp and his colleagues published their paper in 1973, there were perhaps eight known restriction enzymes, They added one by showing that *Haemophilus parainfluenzae*, which had previously been thought to contain a single enzyme, actually had two which acted at differem sites. Richard Roberts had already begun a concerted search for these enzymes in all sorts of bacteria. He 'felt sure they must exist...This was a somewhat heretical view at the time, in that most of my colleagues told me that it was unlikely that there would be more enzymes of this sort.' Roberts's conviction proved correct; a steady stream of new restriction enzymes with new recognition sites followed, and with them a similar' continuous stream of visitors with DNA samples who wished to see which enzymes would cut them'. This led to the first, private catalogue, in 1974, and by the published list of 1976 Roberts had documented more than 80 restriction enzymes. The number now is above the 300 mark.

The plethora of enzymes could have become a source of confusion, but very early on Smith and Nathans, with remarkable prescience, saw where they were heading and polled many of their colleagues in the field to bring some order to it all. Every restriction enzyme would have a specific name which would identify it uniquely. The first three letters, in italics, indicate the biological source of the enzyme, the first letter being the initial of the genus and the second and third the first two letters of the species name. Thus, restriction enzymes from *Escherichia coli* are called *Eco*; *Haemophilus influenzae* becomes *Hin*; *Diplococcus pneumoniae Dpn*; and so on. Then comes a letter that identifies the strain of bacteria: *Eco* R for strain R (strictly for *E. coli* harbouring a plasmid called R), *Eco* B for B. Finally there is a roman numeral for the particular enzyme if there is more than one in the strain in question; *Eco* RI for the first enzyme from *Escherichia coli* R, *Eco* RII for the second.

The enzymes, having been named, can also be divided into three groups. Class I enzymes are the troublesome ones like Meselson. *Eco* K, which recognize a specific sequence but don't cleave the DNA at a specific point: they walk a random distance from the recognition site and break the DNA there. Class II enzymes are the workhorses of the genetic engineer, Like Smith's *Hin* they recognize a specific site and break the DNA at a particular place within that site, A very few class II enzymes are a bit odd; like all restriction enzymes, they recognize a specific sequence, but they don't cleave the DNA within the recognition site; instead, they make a break a set number of nucleotides away. *Mbo* II, for example, from the bacterium *Moraxella bovis* (which causes pinkeye in cattle), recognizes

5′ —G—A—A—G—A— 3′
3′ —C—T—T—C—T—5′

but makes its break some way down stream of this, to create a fragment with

5′ —G—A—A—G—A—N—N—N—N—N—N—N—N— 3′
3′ —C—T—T—C—T—N—N—N—N—N—N—N— 5′

at the end. These 'odd' class II enzymes used to be placed, by European molecular biologists in particular, into a third class, class III. But American workers, and Roberts in particular, adopted a slightly different classification. Their class III contained enzymes like Eco PI, which 'have characteristics intermediate between those of the Type I and Type II restriction endonucleases'. 'The original classification made more sense,' says one European biologist, 'but Roberts has become pre-eminent and so ... we have adopted the current usage'. 'For practical purposes, the random effects of class I and class III restriction enzymes means that they don't find much use in applied genetic engineering. The real stars are the class II enzymes.

What they Do

By making a wise choice from the catalogue of known enzymes, the genetic engineer can find one to perform practically any task. Almost any sequence of bases can be located and cut at will. Some enzymes recognize a long sequence, six or seven bases long: they are often useful for opening a circular strand of DNA at just one point. Others have a much smaller site, four or even three bases long: these will 'produce small fragments that can then be used to determine the sequence of bases along the DNA. Enzymes from different sources often recognize the same site. They are called isoschizomers, and while some cleave at the same place in the site, others cleave at different places. Nobody knows whether isoschizomers are 'the same' enzyme, in terms of their exact structure; it seems unlikely. Enzymes such as Hin dII, the first to be characterized, allow some flexibility within a rigid site and are very useful. So cuts can be made anywhere along the DNA, dividing it into many small fragments or a few longer ones, and in an utterly repeatable fashion. The cuts made by a type II enzyme on a given sort of DNA will always be the same.

Strictly speaking, an enzyme qualifies as a restriction endonuclease only if the bacteria also has a specific modification enzyme to protect its own DNA; but in many cases the restricting activity is all that the genetic engineers are interested in. Nevertheless, some of the modification enzymes have been isolated and characterized, and they inevitably do mirror the recognition site of the restriction enzyme, possibly because the two sorts of enzyme share a subunit that is dedicated to seeking out a particular sequence of bases along the chain. But there are some oddities in the restriction-modification pattern. *Diplococcus pneumoniae* has two enzymes, *Dpn* I and *Dpn* II, that recognize the same four-base site, GATC. *Dpn* II is completely normal, and works like any other type II enzyme, but *Dpn* I will break the strand only if the site is modified by having a methyl group attached to the adenine. This is extremely odd. Why should the normal run of events be upset in this way? And how does *Diplococcus pneumoniae* protect itself from its two potentially devastating restriction enzymes? Perhaps the bacteria don't use this enzyme as a restriction enzyme. As Hamilton Smith says,' it is different to rationalize this reversal of the normal role of methylation.

Like the recognition site, the cut that each enzyme makes varies from enzyme to enzyme. Some, like *Hin* dII make a clean cut straight across the double helix, Fragments from *Hin* dll have ends that are flush. Others make a staggered break, Eco RI, for example, recognizes

5′ –G—A—A—T—T—C— 3′
3′ —C—T—T—A—A—G— 5′

but makes its cut between the G and the A in each strand. This leaves each fragment with a protruding single strand at the 5' end. *Pst* I, from the gut living bacteria *Providencia stuartii*, does the reverse. It recognizes the sequence

5′ —C—T—G—C—A—G— 3′
3′ —G—A—C—G—T—C— 5′

and cuts between the A and G, leaving the protruding end at the 3′ end of the double helix.

Enzymes that make a staggered cut-are especially important, because the single strands that they leave protruding are complementary in base sequence. Under the right conditions, the complementary bases will pair up again, so if you cut two different sorts of DNA with one of these enzymes and mix all the fragments, the chances are that fragments from the two sorts of DNA will come together in a new hybrid molecule. You will have made a new combination of genes - recombinant DNA.

What's it all for?

There are two ways to answer this simple question. One is the answer that occupies much of the rest of this book, that restriction enzymes are 'for' genetic engineering. Certainly it would be impossible without them. But they didn't evolve to help us manipulate DNA. So what, in the evolutionary sense of 'what good', are restriction enzymes really for'?

The obvious answer would seem to be that they are a defence against viral infection. This certainly seems reasonable enough. After all, that is how the phenomenon was first uncovered. Unmodified bacteriophages cannot grow very well in hosts that possess restriction enzymes, so restriction enzymes do protect cells from viruses. But is that all? Probably not, because if restriction really is a, defence against viruses it is a very inefficient one. There is no defence against viruses that carry the modification, and it seems a little far-fetched to say that the whole elaborate mechanism evolved to protect bacteria of one strain from viruses that had most recently grown on a different strain. In any case, a far better evolutionary defence against viruses would be to lose, or modify, the sites on the outside of the bacteria that the virus recognizes and attaches to before inserting its DNA into the host. And if restriction were a protection against infection, we should expect to see far more of it in higher organisms, which are just as prone to viral attack as bacteria. In fact, there still very little evidence of any restriction activity in any eukaryote cell.

So it looks as if restriction did not evolve to protect bacteria from invading phage DNA, although it incidentally serves that purpose now. What is left in the way of invading DNA? DNA from other bacteria. When bacteria mate, they exchange DNA. If different strains were to mate, restriction enzymes in the recipient would usually ensure that the DNA from the donor was from a cell with the same restriction modification complex. The system would act to keep strains pure, by destroying DNA from different strains. The same could be true of higher organisms. A single-celled green alga, *Chlamydomonas*, does have a restriction-modification enzyme system, but the enzymes do not seem to have anything to do with attack by viruses. Instead, they destroy *Chlamydomonas* DNA after mating, The DNA in the chloroplasts from the male partner is not methylated, and is destroyed by the female's restriction enzyme, ensuring that the offspring inherit their chloroplasts exclusively in the maternal line. Salvador Luria, talking about bacteria, says:

The branding-and-rejection system facilitates the evolution of bacterial strains in diverging directions, in the same way that isolation mechanisms in cross-fertilization play a role in the evolution of plant and animal species.

Table 12.1. A sample of restriction enzymes

Source organism	*Abbreviation*	*Recognition and cleavage site (5′ → 3′) (3′ → 5′)*
Bacillus amyloliquefaciens H	Bam HI	G|GATCC CCTAG|G
Escherichia coli RY13	Eco RI	G|AAT|TC CTTAA|G
Haemophilus aegyptius	Hae II	Pu GCGC|Py Py|CGCGPu
Haemophilus aegyptius	Hae III	GG|CC C|GG
Haemophilus haemolyticus	Hha I	GCG|C C|GCG
Haemophilus influenzae R_d	Hin dII	GTPy|PuAC CAPu|Py TG
Haemophilus influenzae R_d	Hin dIII	A|AGCTT TTCGA|A
Haemophilus parainfluenzae	Hpa I	GTT|AAC CAA|TTG
Haemophilus parainfluenzae	Hpa II	C|C|GG GGC|C
Providencia stuartii 164	PstI	CTGCA|G G|ACGTC
Streptomyces albus G	Sal I	G|TCGAC CAGCT|G

More important than the evolution or purity of strains and species is the restriction modification complex itself. By destroying DNA associated with any other modification enzyme, the recognition part of the complex ensures the survival of copies of itself, regardless of what species or strain they may be in.

Restriction enzymes pose fundamental questions, not only about their evolutionary *raison d'eter* but also about the mechanics of how they work. The interaction between protein, the recognition portion of the enzyme, and nucleic acid, the DNA it cleaves, is at the heart of understanding how cells regulate the expression of their genetic material. Not all the genes are being listened to at once, and proteins play a vital role in controlling and orchestrating the operation of the whole genome. Restriction enzymes, as models of more general protein-nucleic acid systems, will doubtless provide fascinating insights to those prepared-to pursue them. But for the vast majority of molecular biologists, restriction enzymes are of importance not because they might provide fundamental insights or anything like that: they cut DNA, and that's what counts.

13

Molecular Fluctuation

There are now many pathological examples of the deletion, insertion, duplication and expansion of human genes causing inherited disease. Similar mutations have however also occurred over evolutionary time. Far from being invariably disadvantageous, such mutational changes have often been recruited by the opportunistic evolutionary process and now contribute to both gene and genome architecture. These types of mutation have led to significant changes in gene size and number in different lineages and their contribution to the evolution of extant human genes will now be reviewed.

Gross Gene Deletions in Evolution

Gross Gene Deletions during Primate Evolution

Gross gene deletions may arise through a number of different recombinational mechanisms but probably the most common is likely to be *homologous unequal recombination* (occurring either between related gene sequences or between repetitive elements). Thus, *Alu* sequences flanking deletion breakpoints have been noted in a considerable number of human genetic conditions and may represent favoured sites for recombination and hotspots for gene deletion. Chromosomally duplicated regions (*duplicons*) are often common sites for pathological rearrangements, particularly gross deletions, since they have the potential to mediate homologous unequal recombination events (e.g. 15q11-q13). Not surprisingly, several instances of gross gene deletion have been noted during primate evolution. One such example is the loss of the γ1-globin gene in New World monkeys (with the notable exception of the capuchin monkey, *Cebus albifrons*) due to a 1.8 kb deletion which has removed most of exon 2, all of intron 2, exon 3, and much of the 3' flanking region. As a result, γ2-globin is the primary fetally expressed globin gene in New World monkeys whereas in Old World monkeys, it is γ1-globin.

Another example of gross gene deletion in primates is the loss of one of the haptoglobin (HPP; 16q22.2) genes in humans which occurred, probably via homologous unequal recombination, after the separation of the human and chimpanzee lineages. Finally, although there are two semenogelin genes (SEMG1, SEMG2; 20q12-q13.1) in humans and most old world and New World monkeys, the *Semg2* gene has been deleted from the genome of the cotton-top tamarin (*Saguinus oedipus*). There is no evidence for any selective advantage resulting from any of these gene deletions. In all cases cited, a similar paralogous gene was available to substitute for the deleted locus. This genetic redundancy probably ensured that, owing to the absence of purifying selection, such deletions came to be fixed through genetic drift.

Gross Deletional Polymorphisms

Gross gene deletions in two distinct glutathione S-transferase genes have been found to occur as polymorphic variants in various human populations. In humans, four gene families of glutathione S-transferases encode a series of enzymes responsible for the metabolism of a wide range of xenobiotics. The gene for the mu-class glutathione S-transferase (GSTM1; 1p13.3) is absent from between 10% and 64% of individuals depending upon the population under study. Loss of the GSTM1 gene is due to a 15 kb deletion probably brought about by homologous unequal recombination between two almost identical 4.2 kb repeats that flank the GSTM1 gene. These repeats are likely to have originated with the original duplication that gave rise to the mu-class glutathione S-transferase genes more than 20 Myrs ago. A similar deletional polymorphism also occurs in the theta-class glutathione S-transferase (GSTT1; 22q11.2) gene; the gene is absent in ~38% of the Caucasian population. Since the GSTM1 and GSTT1 genes are involved in

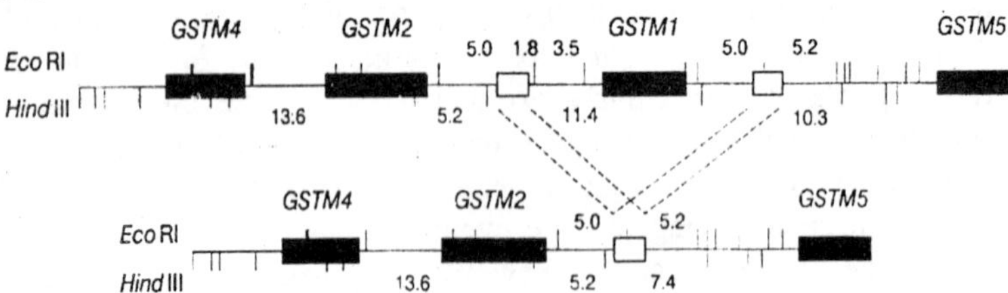

Fig. 13.1. Model for homologous recombination between 4.2 kb repeats (open boxes) flanking the human mu-class glutathione S-transferase (GSTM1) gene leading to gene deletion.

xenobiotic metabolism, it is quite possible that the presence/absence of these polymorphic alleles are not selectively neutral.

Dichromacy (colour blindness) is very common (occurring at a polymorphic frequency (5-8%) in Caucasian males) with individuals so affected having either a deletion of the green colour pigment (GCP) gene or possessing a hybrid red/green colour pigment (RCP/GCP) gene in its place. Further variable gene number polymorphisms, which probably also arose by homologous unequal recombination, are apparent in the human α-amylase gene cluster at chromosome 1p21 and the pepsinogen A gene cluster at 11q13. In the first case, a 'short' haplotype lacks AMY1A, AMY1B and the pseudogene AMYP1 whilst a 'long' haplotype contain two extra copies of duplicated fragment containing AMY1A, AMY1B, and AMYP1. In the second case, the three most common haplotypes were PGA-A (containing the PAG3, PGA4, and PGA5 genes), PGA-B (containing the PGA3 and PGA4 genes) and PGA-C (containing only the PGA4 gene). Copy number polymorphisms due to gene deletions have also been reported at the human T-cell receptor β (TCRB; 7q35) and γ (TCRG; 7p15) loci, the ζ-globin (HBZ; 16p13.3) gene, the rhesus blood group D antigen (RHD; 1p34–36.2) gene, the immunoglobulin heavy chain constant region γ4 (IGHG4; 14q32) gene and the complement C4A (C4A; 6p21.3) and C4B (C4B; 6p21.3) genes. In some gene clusters, it can be difficult to ascertain whether polymorphic alleles have arisen by gene deletion or duplication.

Microdeletions in Evolution

It is possible to extrapolate from lessons learned through the study of microdeletions in a pathological context to microdeletions that have occurred during gene evolution? In particular, can we gain insight into the nature of the generative mechanism(s) underlying evolutionarily significant microdeletions and the possible influence of the local DNA sequence environment? Although in principle the answer to this question is likely to be in the affirmative, in practice the DNA sequences that were originally responsible for mediating the microdeletions have often decayed or been lost and it may not always be possible to reconstruct them.

Microdeletions in Pathology

Microdeletions (<20 bp) causing human genetic disease were analyzed by Cooper and Krawczak (1993) in an attempt to relate the presence of specific DNA sequence motifs in the vicinity of these lesions to possible mechanisms responsible for their generation. In many cases, slipped mispairing at the replication fork between homologous sequences in close proximity to one another on complementary DNA strands appeared to be the causative mechanism. Slipped mispairing probably occurred either between *direct repeats* or through the formation of secondary structure intermediates potentiated by the presence of *inverted repeats* or *symmetric elements*. A consensus sequence, TGRRKM, common to pathological deletion hotspots has been noted in a number of different human genes. This *deletion hotspot consensus sequence* is similar to the core motifs, TGGGG and TGAGC, found in immunoglobulin switch ($S\mu$) region and to putative arrest sites for DNA polymerase α. Cooper and Krawczak (1993) also found that a second motif (polypyrimidine runs of at least 5 bp; YYYYY) was over-represented in the vicinity of short human gene deletions whilst Monnat et al. (1992) observed a significant association between HPRT1 (Xq26.1) gene deletion breakpoints and CTY vertebrate topoisomerase I cleavage sites. In principle, such sequence motifs may also have promoted the occurrence of microdeletions during evolution. Indeed, in probably the large study of its kind to date, a sequence comparison of orthologous and paralogous members of the primate T-cell receptor β (TCRB; 7q35) gene family, micro-deletion breakpoints appear to be frequently flanked by polypyrimidine runs and sequences which possess marked homology to the deletion hotspot consensus sequence.

Microdeletions Mediated by Direct Repeats

Micro-deletions occur during gene evolution at a frequency 10% that of nucleotide substitutions. Pairwise comparisons of the noncoding regions of human, rabbit and murine β-globin genes have shown that they differ from each other in terms of numerous deletions/insertions and Efstratiadis et al. (1980) proposed that the short direct repeat (2-8 bp) sequences immediately flanking these sites could have templated the generation of these lesions by slipped mispairing. Direct repeats may also have been involved in generating the two inactivating single base-pair deletions (del C822 and del G904) noted in the human α-1,3 galactosyltransferase (GGTA1; 9q34) gene. Since chimpanzees possess both these deletions, whereas orangutan and gorilla only have del904, it would appear that del904 was the original inactivating mutation. C822 is immediately flanked by imperfect direct repeats (TACAGGC$\underline{C}$T and TACAAGGCAG, where $\underline{C}$ is nucleotide 822) that could have templated the 1 bp deletion via slipped mispairing. Slipped mispairing may also have been responsible for the G904 deletion since G904 is the 3' most base of a string of five Gs.

```
5' FLANKING
Human  δ      CCA-------GCATAAAA
Human  β      CCAGGGCTGGGCATAAAA
LARGE INTRON
Human  Aγ     ATGACTTTT----ATTAGAT
Human  Aγ     ATGACTTTTCTTTATTAGAT
Human  Aγ     TGTGTGTGTGTG-------------------TGTGTGTGTGTG
Human  Gγ     TGTGTGTGTGTGTGTGTGTGCGCGCGTGTGTTTGTGTGTGTGTG
Rabbit β      AAGTA----------------CTTTCTCTAATC
Mouse  β      AGTCCTTCTCTCTCTCCTCTCTCTTTCTCTAATC
Rabbit β      TGGTAG--------------------------------AAACAACT
Mouse  β      TTGGCTTTTATGCCAGGGTGACAGGGGAAGAATATATTTTACATAT
Mouse  βmaj   TGACATAGG----------------------------------------ATTCT
Mouse  βmin   TGTCATAGAATAATTCTTTTTTATTTTTATTTATTATTTTTTTCATAGAATAATTCT
Mouse  βmaj   TGTGTGTGG------------AGTGTT
Mouse  βmin   GGTGTGTGGATGTGAATTGTGAGTGTT
3' NONCODING
Human  ε      CAGGT-----------GTTCCT
Human  β      CCAATTTCTATTAAAGGTTCCT
Human  Aγ     GCAATACAAA--------------------------------------------TAATAAA
Human  β      GTCCAACTACTAAACTGGGGGATATTATGAAGGGCCTTGAGCATCTGGATTCTGCCTAATAAA
Human  Aγ     ATACAA-----------------------------------------TAATAAA
Human  δ      TATTTTCTGAACTTGGGAACAATGAATACTTCAAGGGTATGGCTTCTGCCTAATA A
Rabbit β      AAAAATTAT----------------------------------GGGGACA
Human  β      AATTTCTATTAAAGGTTAATTTGTTCCCTAAGTCCAACTACTAAACTGGGGGATA
```

Fig. 13.2. Examples of deletions flanked by short direct repeats within non-coding sequences of mammalian β-globin genes.

A direct repeat may also have templated the single base deletion in the 5' flanking region of the human interferon α10 (IFNA10; 9p22) gene relative to the other α-interferon genes. Flanking direct AGGT repeats appear to have mediated an AGG deletion exhibited by both orthologous and paralogous members of the primate T-cell receptor β (TCRB; 7q35) gene family whilst in the same gene family, overlapping 7 bp direct repeats (CTTTTCTTTTCT) may have served to template a TTTCT deletion.

Microdeletions Mediated by Inverted Repeats

Inverted repeats may also have mediated the generation of microdeletions during gene evolution. One example is the inactivating 13 bp deletion in exon 2 of the gibbon urate oxidase gene; two imperfect inverted repeats (CAAGAAC and GTTCATG) span the breakpoints of this deletion. A 20 bp deletion has been reported from the 5' region of the δ-globin (Hbd) gene of the colobus monkey, *Colobus polykomos*. This deletion, which spans the transcriptional initiation site used in Old World monkeys and anthropoid apes, is responsible for a five-fold reduction in *Hbd* gene transcription as assessed by *in vitro* transcription assay. Inspection of the putative deleted bases and the flanking DNA sequence reveals the presence of a 13 bp imperfect

```
                              +1
C aagggagggcagag------------------CTTCTGA
R       g  a     a ccaactgttgcttATACTTG
B       g  g     a tcaactgttgcttACATTTG
S       c  g     g ctaactgttgcttTGACTTG
H       c  g     a tcgactgttgcttACACTTT
```

Fig. 13.3. Alignment of δ-globin (Hbd) gene sequences from colobus monkey (C), rhesus macaque (R), baboon (B), spider monkey (S) and human (H) showing the location of a 20 bp deletion in the colobus monkey.

inverted repeat which could have been responsible for the deletion through formation of a hairpin loop.

An inverted repeat also appears to have templated the deletion of a GAT codon in the human interferon α2 (IFNA2; 9p22) gene relative to the other α-interferon genes. Finally a contiguous inverted repeat sequence (ATTC-CCAGTTTCTGGGAAT) may well have templated an 8 bp deletion exhibited by both orthologous and paralogous members of the primate T-cell receptor β (TCRB; 7q35) gene family.

Microdeletions in Vertebrate Evolution

The comparison of gene/protein sequences between humans and the great apes also yields examples of *in-frame* microdeletions that must have occurred during primate evolution. Thus, amino acid residue Glu9 of the blue cone pigment protein (BCP; 7q31-q35) present in the talapoin monkey *Miopithecus talapoin* (an Old World primate) and in the marmoset *Callithrix jacchus* (a New World primate) is absent from the human protein and appears to have been deleted from the BCP gene within the human lineage. The functional consequences of the removal of this amino acid residue are however unclear.

Some gene region harbor a disproportionate number of deletions/insertions inferred from alignment gaps noted in sequence comparisons e.g. exons 6 of the orthologous amelogenin (AMELX, Xp22.1-q22.3; AMELY, Yp11.2) genes of various vertebrates. These lesions have dramatically reduced the similarly between vertebrate. These lesions have dramatically reduced the similarity between vertebrate amelogenins in the Pro/Gln-rich region of the protein as compared with that manifested by other region. Some microdeletions occurring during evolution may have been advantageous by virtue of their alteration of a protein product, others through a change in the reading frame bringing about gene inactivation. Of course, micro-deletions need not necessarily have conferred any selective advantage; even if merely neutral with respect to fitness, they could have become fixed by genetic drift alone.

Microinsertions in Evolution

Microinsertions that have occurred during evolution have scarcely been studied. However, the underlying generative mechanisms are likely to be broadly similar to those causing human genetic disease. In their study of microinsertions in human genes causing inherited disease, Cooper and Krawczak (1991) concluded that insertional mutation involving the introduction of <10 bp DNA sequence into a gene coding region was not a random process and appeared to be highly dependent upon the local DNA sequence context. Further, mechanistic models which have explanatory value in the context of gene deletions were found to be useful in accounting for the nature and location of gene insertions.

In noncoding DNA, insertions are about half as frequent as deletions and mostly involve single nucleotides. The rate of gap formation, regardless of whether caused by insertions or deletions, has been estimated to be $\sim$0.15-0.17 kb^{-1} $Myrs^{-1}$. In practice, studies of the DNA sequence environment of microinsertions that have occurred during evolution are likely to be rather difficult since the original sequence context of the insertion or deletion will often have become obscured by subsequent mutation.

Gene Coding Region Microinsertions

Microinsertions occurring during human gene evolution can be found by sequence comparison of either orthologous or paralogous genes/proteins. Thus, the 54 bp insertion in the promoter of the human liver arginase (ARG1; 6q22.3-q23.1) gene is absent in the orthologous gene of *Macaca fascicularis*. Similarly, a 37 bp insertion has been introduced into the promoter of the orthologous Duchenne muscular dystrophy (DMD; Xp21) gene of the spider monkey, *Ateles geoffroy*; the inserted sequence is flanked by two TAAA repeats. A 12 bp insertion in the T-cell receptor α-chain (TCRA; 14q11.2) gene encodes an Ile-Pro-Ala-Asp tetrapeptide (residues 88-91) that is specific to the primate lineage. Interestingly, the region of the T-cell receptor α-chain protein between positions 86 and 91 appears to be a hotspot for insertional events during evolution: the rabbit and rat genes appear to have acquired a 3 bp (single amino acid) insertion, whereas the bovine, ovine, and murine genes manifest a 6 bp (double amino acid) insertion at this position. Finally, a 24 bp sequence found in the transmembrane domain region of the human glycophorin E (GYPE; 4q28-q31) gene appears to have been derived from the paralogous glycophorin B (GYPB; 4q28-q31) gene during primate evolution prior to the divergence of the gorilla from the lineage of the other great apes. It is unclear whether this insertion event was mediated by homologous unequal recombination or gene conversion.

Microinsertion Polymorphisms

Several microinsertion polymorphisms have been reported in human genes. Thus, a single nucleotide insertion polymorphism has been noted in the promoter region of the insulin promoter factor 1 (IPF1; 13q12.1) gene. A single nucleotide insertion polymorphism is also present in the ABO blood group (ABO; 9q34) gene which serves to inactivate it. Finally, a 9 bp insertion polymorphism in exon 9 P450 CYP2D6 (22q13.1) gene occurs in the Japanese population and is associated with a poor metabolizer phenotype

Indels

Clearly, in extant proteins, selection must have ensured the retention of essential features of tertiary structure. Indeed, insertions or deletions which altered the reading frame must have been rendered harmless in order for the protein to retain its biological activity. Thus, the insertion of bases inferred in one member of a paralogous protein pair often implies a counterbalancing deletion in the immediate vicinity (or *vice versa*) to restore the reading frame. Such 'indels' tend to involve sequences of between 1 bp and 5 bp in length. They are generally found in turn and coil structures and rarely interrupt α-helices and strands. One example of a simple indel occurring during evolution is the deletion of an AA doublet and its replacement with a GT doublet in the 5' flanking region of the human interferon "α9" gene.

An example of a more complex indel that has occurred during vertebrate evolution is provided by the human chorionic gonadotropin β-subunit (CGB; 19q13.3) gene and is responsible for the introduction of a 24 amino acid C-terminal extension to the protein product. The CGB gene emerged as a result of the duplication of an ancestral β-luteinizing hormone-like (LHB; 19q13.3) gene. A single base deletion (A1540) altered the translational reading frame of the ancestral gene allowing read-through into what was originally 3' untranslated region (UTR). An insertion of a CG dinucleotide also occurred at nucleotide 1612 within the ancestral 3' UTR region and served to extend the chorionic gonadotropin β-subunit protein by a further 8 amino acids. The CGB gene therefore evolved from its LHB-like ancestor by acquiring 8 amino acids through translation in a new reading frame and incorporating 24 novel amino acids from the 3' UTR into its coding sequence.

Insertion of Transposable Elements in Evolution

Transposable elements in the human genome are essentially of two kinds, those that undergo transposition through a DNA intermediate. Transposable elements with a DNA intermediate are termed *transposons* and these are characterized by terminal inverted repeats and duplication of the target site (visible as direct repeats flanking the element). However, the great majority of transposable elements in the human genome have undergone retrotransposition through an RNA intermediate. Such *retroelements* or *retroposons* may be of either viral or nonviral origin. Whilst endogenous retroviral sequences comprise some 0.1-0.6% of the human genome, the nonviral *Alu* sequences and LINE elements may together make up as much as 10% of the genome.

Endogenous Retroviral Sequences and Transposable Elements

Retroposons

A number of retroposon families have been characterized in primate genomes. Occasionally, these are human-specific (e.g., the HERV-K10 related SINE-R.C2) but usually they are found distributed through the genomes of other primate species (e.g. RTVL-H, RTVL-I, HERV-K, and HERV-L. Type I and II HERV-H elements were amplified to ~1000 copies after the divergence of New World from Old World monkeys but before the divergence of apes from Old World monkeys. By contrast, the family of type Ia HERV-H elements expanded to ~100 copies only after the divergence of apes from Old World monkeys. Analysis of the copy number, distribution and sequence characteristics of such endogenous elements promises to provide important clues as to the evolutionary history and phylogeny of the various mammalian orders, suborders, and species, and even the population genetics of human racial groups.

Retroposons have sometimes become integrated into the vicinity of human genes. Thus, two copies of an RTV_L-I sequence are present in the human haptoglobin (HP; 16q22.1) gene cluster whilst an additional copy has been inserted in the same region in the orthologous chimpanzee gene. Rather more dramatically, the endogenous retrovirus, HRES-1 lies within the coding sequence of the human transaldolase gene (TALDO1; 11p15).

One view of endogenous retroviral elements is that they inserted themselves into the germline of our primate ancestors during the last 40 Myrs as a result of infection with exogenous retroviruses, persisting thereafter as proviruses, albeit rendered replication-defective by multiple

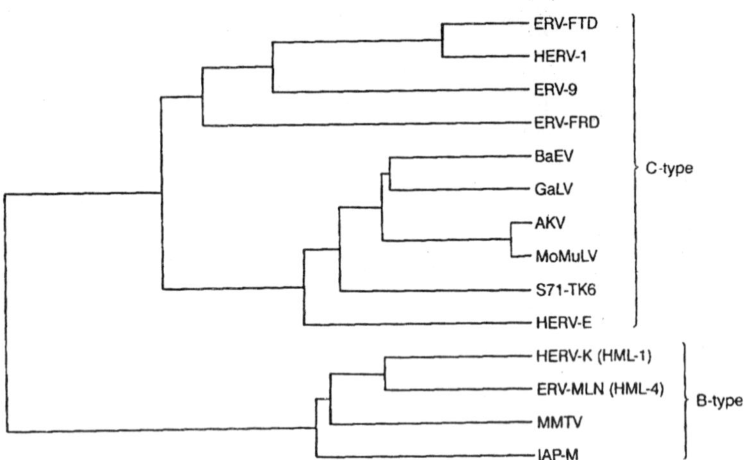

Fig. 13.4. Phylogenetic analysis of mammalian endogenous retroviral pol sequences.

mutational events. Another (not incompatible) view is that retroviruses themselves originally arose from intracellular retroelements (the protovirus hypothesis), a view which is supported by the phylogenetic analysis of endogenous retroviral DNA sequences. Regardless of whether or not the horizontal transmission of retroviral elements has taken place, copy number amplification has certainly occurred.

Retrotransposition of generates retroposon sequence variants owing to its inherent imprecision: target site rearrangements combine with the infidelity of both reverse transcriptase and RNA-dependent RNA polymerases to ensure that the inserted sequences are highly variable thus providing new avenues for the highly opportunistic evolutionary process.

Transposons

Transposon-like THE-1 repeats, which lack any obvious homology to retroviral sequences, have been found in the deletion-prone intron 43 of the dystrophin (DMD; Xp21.2-p21.3) gene, the human blood group GC (4q12) gene and the 3' untranslated region of the human calmodulin-protein (CALML1; 7p13-pter) gene. A cluster of three THE-1 repeats located in a 26 kb region of intron 7 in the human DMD gene has arisen by three independent insertion events. There is some evidence to support the hypothesis that the insertion of these elements has occurred at preferred target sites.

A pseudoautosomal gene sequence (*Tramp*) has recently been isolated which encodes within its single exon a protein with homology to transposases (enzymes that mediate transposition) of the Ac family. It is as yet unclear whether the Tramp protein has been involved in the transposition of other transposable elements or if it has instead become specialized for a novel cellular function. The centromeric protein CENP-B (CENPB; 20p13) may also represent an example of a transposase-encoded protein which has acquired a cellular function. This protein binds to the CENP-B box (TTCGNNNNANNCGGG) sequence in the alpha satellite DNA of human centromeres and has sequence similarity to the *pogo* family of transposases which includes the *Tiger* elements. Since CENP-B has nicking activity, it may promote homologous recombination and could have contributed to the species-specific patterns of evolution of satellite repeats.

Since many transposable elements contain enhancer sequences, their transposition may have served to alter the pattern of host gene expression at or around the integration site. Thus, once transposed, evolution may have recruited such enhancers to play a role in the transcriptional regulation of a gene in the vicinity of the integration site. One example of this is the human salivary amylase (AMY1C; 1p21) gene where the HERV-E-derived enhancer may be involved in tissue-specific expression.

Evolution may also recruit transposable elements as a means to alter mRNA processing. For example, a B2 (SINE) element has become inserted into the 3' untranslated region of the murine (*lifr*) gene encoding the soluble form of the leukemia inhibitory factor receptor (LIFR). Insertion of the B2 element has, by potentiating alternative 3' mRNA processing and alternative splicing, given rise to a truncated mRNA species (relative to the mRNA encoding the membrane-anchored LIFR) which encodes soluble LIFR. In the rat, no such retrotranspositional event has occurred and the soluble form of LIFR is not found.

A very special case of the opportunistic recruitment of a transposable element may have been that of the recombination-activating gene (RAG) transposase postulated to have been inserted into an ancestral immunoglobulin/T-cell receptor gene soon after the divergence of jawed and jawless fishes. The subsequent conversion of this transposon into a site-specific recombinase may have been the critical event in allowing the vertebrates to generate the genetic diversity so essential for the flexible adaptive response of their immune systems.

LINE Elements

LINE elements have assumed considerable importance in the context both of gene pathology and gene evolution. They are present in a wide range of mammals and are represented by some 40 subfamilies. They are nonrandomly distributed in the human genome, inserting preferentially into chromosomal G bands and at the DNA level, into A-rich sequences. The total numbers of LINE elements in four of the great apes have been estimated by Hwu et al. (1986): human, 107,000; chimpanzee, 51,000; gorilla, 64,000; and orangutan, 84,000. Since these figures differ markedly, it follows that numerous insertions and deletions of these sequences must have occurred during the evolution of the great apes.

In a pathological context, a number of examples of gene inactivation through insertion of LINE elements into a gene coding sequences are known and in some cases a preference for integration at AT-rich sequences is exhibited. Further, the target sites of two LINE elements inserted into the factor VIII (F8C; Xq28) gene causing hemophilia A are 80% homologous to a 10 nucleotide motif (GAAGACATAC) present in one of the highly favoured retroviral insertion target sequences reported by Smith et al. (1988).

Preferential target sites for LINE elements are also apparent in mammalian genes during evolution. For example, the interleukin-6 genes of rodents represent hotspots for LINE element retrotransposition. During mammalian evolution, the introduction of LINE elements in the vicinity of genes has sometimes altered gene expression as a consequence of their being recruited to perform a regulatory function. LINE elements have served to promote genetic rearrangements and indeed they may well have mediated both gene inversion and duplication events during evolution. Finally, and perhaps most importantly, LINE elements may have repeatedly transduced exons from one genomic location to another, thereby potentiating exon shuffling, the transfer of exons encoding specific protein modules between genes.

Alu Sequences

Evolution of Alu sequences

The fossil *Alu* monomer is thought to have arisen by the deletion of the central S domain of 7SL RNA (RN7SL) followed by the addition of a 3' poly(A) tract which may have facilitated reverse transcription of these RNA polymerase III transcripts. The free left arm monomer then arose by deletion of 42 bp from the fossil *Alu* monomer whilst the free right arm monomer arose by deletion of 11 bp from the fossil *Alu* monomer. The first *Alu* sequence may have been formed by dimerization of a free left arm monomer with a free right arm

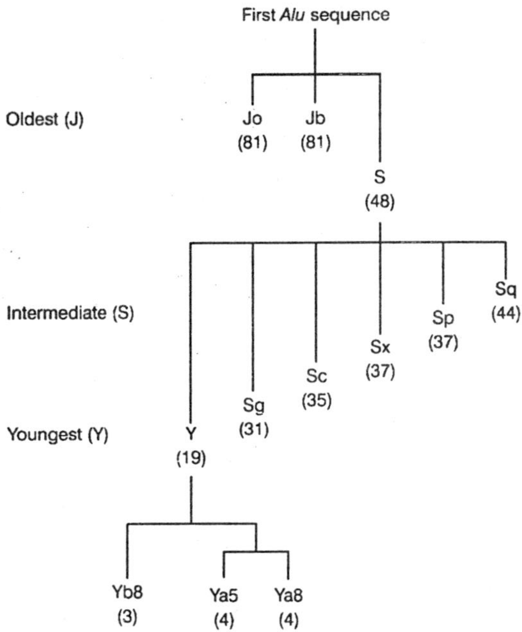

Fig. 13.5. The proposed evolution of the 12 human Alu subfamilies.

monomer an event which is thought to have occurred about 60 Myrs ago, before the divergence of prosimians. Subsequently, many rounds of sequential amplification took place to generate the 12 human *Alu* subfamilies seen today.

The total numbers of copies of Alu sequences in four of the great apes have been estimated by Hwu et al. (1986): human, 910,000; chimpanzee, 330,000; gorilla, 410,000; and orangutan, 580,000. As with the LINE elements, it would appear that numerous insertions and deletions of these sequences have occurred during the evolution of the great apes. At the chromosomal level, *Alu* sequences insert preferentially into R bands whereas at the DNA level, they preferentially integrate into A-rich sequences. During mammalian evolution, the introduction of *Alu* sequences in the vicinity of genes has sometimes altered gene expression as a consequence of their being recruited to perform a regulatory function. *Alu* sequences may also have been involved in, or mediated, many other different types of gene rearrangement during gene evolution including gross deletions, duplications, transpositions, gene fusions, recombination, and gene conversion events.

Alu sequence polymorphisms

Once inserted, specific *Alu* sequences have often been relatively stable in terms of their location during primate evolution. This notwithstanding, some human *Alu* sequences are polymorphic in terms of their presence or absence, a situation which is sometimes also found in other primates. Some of these polymorphisms may be of functional significance e.g. a common insertion/deletion polymorphism (0.4/0.60) within intron 16 of the human angiotensin I converting enzyme (DCP1; 17q23) gene, explicable in terms of the presence or absence of a 287 bp *Alu* repeat, is known to have an important influence on serum enzyme concentration. *Alu* sequence retrotransposition is also an occasional cause of genetic disease.

Alu sequence target sites

Although *Alu* sequences occur on average every 3-6 kb, there are several examples of regions that appear to be preferential target sites for *Alu* sequence insertion in mammalian genes. Thus, a 40 kb region, spanning the spermatid-specific protamine genes PRM1, PRM2 and the transition protein (TNP2) gene (16p13.2) contains a total of 42 *Alu* sequences. Similarly, a 22 kb region telomeric to the HLA-B associated transcript 2 (BAT2; D6S51E; 6p21.3) gene in the HLA class III locus contains 42 *Alu* repeats, whilst a 2.2 kb segment 5' to the human lysozyme (LYZ; 12) gene contains four such repeats.

Alu sequences within protein-coding sequences

Many *Alu* sequences are found within introns and therefore this repeat is represented in heterogeneous nuclear RNA. *Alu* sequences are however also found at different locations within mRNA-homologous sequences, the majority occurring within the untranslated regions (UTRs). Thus, Yulug et al. (1995) reported that 5% of full-length human cDNAs contained an *Alu* sequence, with 82% and 14% of these being located in the 3' UTR and 5' UTR, respectively. In a few cases, however, *Alu* sequences have been incorporated into the coding sequences of human genes and have therefore altered the amino acid sequences of the encoded proteins. Thus, a 279 bp *Alu* sequence spans 103 bp of the coding region of a zinc finger protein (ZNF91; 19p12) gene and extends 166 bp into the 3' UTR. Similarly, 110 bp of *Alu* sequence lies within the coding a region of the lectin-like type II integral membrane protein (KLRC1; 12) gene with 43 bp extending into the 3' UTR. An *Alu* sequence is entirely contained within the coding region of the protein serine/threonine kinase stk2 (STK2; 3p21.1) (279 bp *Alu*) gene. Finally, and perhaps most dramatically, two *Alu* sequences (both with poly(A) tails) are entirely contained within the coding region of the regulato. of mitotic spindle assembly 1 (RMSA1; 17p11.2-p12) gene accounting for 111 amino acids of its coding potential, some 40% of the total.

Splice-mediated insertion of Alu sequences

Alu sequences have also been found to alter protein coding sequences through the *splice-mediated insertion* of the repeat and this probably represents the major mechanism by which *Alu* sequences have entered protein coding regions. The principle involved by reference to the pathological example of ornithine δ-aminotransferase deficiency caused by a single base-pair substitution in an intronic *Alu* element in the human ornithine δ-aminotransferase (OAT; 10q26) gene; this lesion activates a cryptic donor splice site which results in the incorporation of the *Alu* sequence into the mRNA.

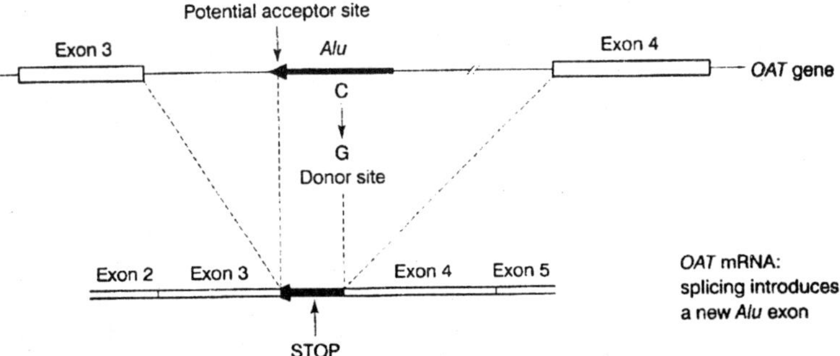

Fig. 13.6. Ornithine δ-aminotransferase deficiency caused by a mutation in a resident intronic Alu element within the OAT gene.

In an evolutionary context, there are several examples of the splice-mediated insertion of *Alu* repeats into human gene coding sequences. The splice-mediated insertion of a 95 bp *Alu* sequence has been reported in the lecithin: cholesterol acyltransferase (LCAT; 16q22.1) gene. In humans, the alternate *Alu*-containing transcript represents between 5% and 20% of the LCAT mRNA. It is also present in LCAT mRNA from chimpanzee, gorilla and orangutan; in the latter, the *Alu*-containing mRNA species constitutes 50% of the total LCAT mRNA pool. It into however present in the LCAT genes of gibbons, or Old World and New World monkeys.

In the human biliary glycoprotein (BGP; 19q 13.2) gene, three mRNA variants are produced as a result of the alternative splicing of an exon (IIa) with one of two virtually identical *Alu* cassettes derived from two intronic repeats. Other such examples are to be found in the

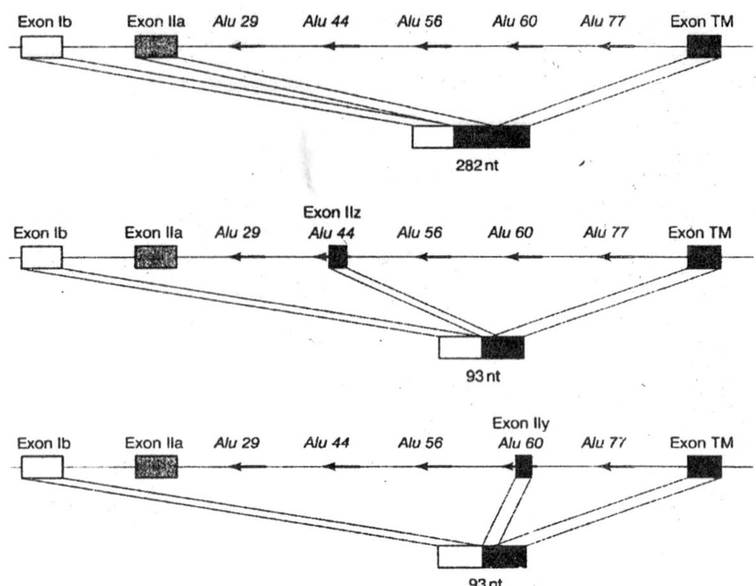

Fig. 13.7. A scheme for alternative splicing in the human biliary glycoprotein (BGP) mRNA.

human REL (2p12-p13) proto-oncogene, and the complement decay-accelerating factor (DAF; 1q32) and complement C5 (C5; 9q33) genes. Of the 17 *Alu* sequences found in mRNA coding regions by Makalowski et al. (1994), seven contained in-frame Stop codons and three others were predicted to cause frameshifts. Thus it is perhaps not surprising that in several cases of mRNA containing *Alu* sequences, *allelic exclusion* is evident and the mRNA containing the *Alu* sequence is of low abundance compared to splice variants of the same gene that lack the repeat. This notwithstanding, it may well be that the splice-mediated insertion of *Alu* repeats has been an important evolutionary mechanism for creating diversity at the protein level.

Alu sequence incorporation by intron sliding

An alternative mechanism of *Alu* sequence incorporation into gene coding regions is intron sliding, and is illustrated by the example of the human HLA-DRB1 (6p21.3) gene; an intronic *Alu* sequence has been incorporated into exon 4 of the HLA-DR-β1 mRNA. Among three variants of the HLA-DR-β1 cDNA, detected by library screening, one was considered to be the usual form, whereas two others were alternatively spliced owing to a lack of splicing at the intron 5 donor site. As a result, exon 5 was extended into a nearby downstream *Alu*

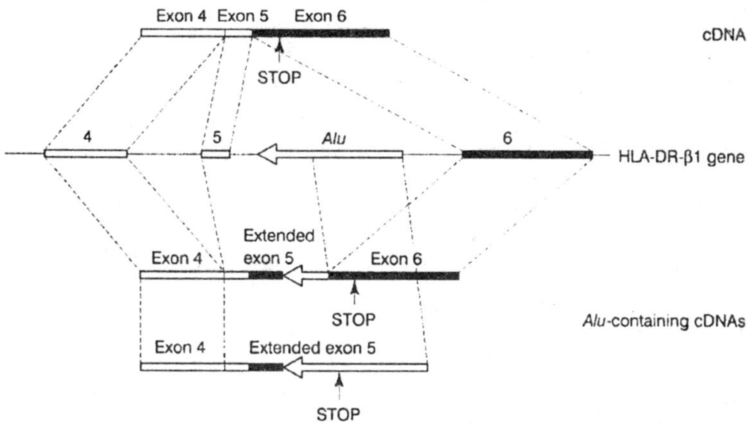

Fig. 13.8. A resident intronic Alu sequence is incorporated into exon 4 of the HLA-DR-β1 mRNA by intron sliding.

sequence in intron 5, either to include a stop codon within the *Alu* sequence, or to be spliced with exon 6 (the open reading frame in the extended exon 5 matches that of exon 6). These three cDNA clones may thus illustrate two phases of intron sliding: the inactivation of an existing splice site followed by the activation of a cryptic one.

Gross Gene Duplications in Evolution

Gene duplication (or partial duplication) events are a fairly uncommon cause of human genetic disease. Two distinct mechanisms are currently envisaged: (i) *homologous unequal recombination* either between homologous chromosome or sister chromatids and (ii) *nonhomologous recombination* at sites with minimal homology. Topoisomerase cleavage sites have been reported to be associated with pathological gene duplications and potential sites for topoisomerases I and /or II have been found to coincide with the breakpoints of duplication in the human factor VIII (F8C; Xq28) and dystrophin (DMD; Xp21.2-p21.3) genes. These observations are potentially interesting since topoisomerase activity has been implicated in several cases of nonhomologous recombination.

One of the best studied gross duplications in human genome pathology is the 1.5 Mb duplication of the short arm of chromosome 17 associated with Charcot-Marie-Tooth disease type 1A. This recurring duplication is thought to be mediated by homologous unequal recombination between two misaligned ~30 kb CMT1A-REP repeat sequences flanking the CMT1A region in direct tandem orientation. In humans, these repeats are 98% identical. Chimpanzees have two copies of a CMT1A-REP-like sequence, whereas gorilla, orangutan and gibbon only have a single copy consistent with a duplication of the CMT1A-REP sequence after gorilla diverged from the human lineage but before the divergence of chimpanzee and human. Orthologous sequence comparison has provided evidence that the distal repeat was the progenitor copy.

Duplications and the Emergence of Paralogous Genes

During vertebrae evolution, novel genes have arisen by genome duplication intra-chromosomal region duplication, and localized individual gene duplication. All three mechanisms give rise to *paralogous* genes, genes that occur within the same species and which have a common ancestor. Paralogous genes therefore include the members of multigene families and superfamilies. Evidence for the common ancestry of paralogous genes may come from sequence homologies (e.g. as with the voltage-sensitive ion channel genes) and/or from similar exon-intron organization, for example the cholesterol ester transfer protein (CETP; 16q21) and the phospholipid transfer protein (PLTP; 20q12-q13) genes, or the growth hormone receptor (GHR; 5p12-p14), prolactin receptor (PRLR; 5913-p14) and interferon receptor α, β, and ω, 1 (IFNAR1; 21q22.1) genes.

Intra-chromosomal Regional Duplication

In the human genome, whole chromosomal segments have sometimes been duplicated resulting in a series of paralogous genes retaining their syntenic arrangement. Thus, a number of genes located at 6p21.3 have paralogous genes at 9q33-q34; the chromosome 6 loci include genes for type 11 collagen α2 subunit, (COL11A2), NOTCH4, 70 kDa heat shock proteins (HSPA1A, HSPA1B, HSPA1L), valyl-tRNA synthetase 2 (VARS2), complement components (C2, C4A, C4B), pre B cell leukemia transcription factor 2 (PBX2) and retinoid X receptor β (RXRB) whilst the chromosome 9 paralogues include COL5A1, NOTCH1, HSPA5, VARS1, C5, PBX3, and RXRA. Other extensive chromosomal duplications included genomic segments present at Xq28 and 16p 11.1 involving the paralogous creatine transporter genes SLC6A8 and SLC6A10 respectively. Some intra-chromosomal duplications may only involve one or a small number of genes, for example the duplication of the iduronate-2-sulphatase (IDS) locus at Xq28. Another example is that of the inverted duplication at 5q13 which duplicated the spinal muscular atrophy (SMA) gene, the survival motor neuron (SMN) gene and the apoptosis inhibitory protein (NAIP) gene. Duplicated paralogous genes may however be translocated to quite different locations on the same chromosome, for example the adrenergic receptor (ADRA1B and ADRB2) genes on 5q23-q32 are quite distant from the evolutionarily related serotonin receptor (HTR1A) gene on 5cen-q11.

Tandem Duplications

Multigene families often form syntenic gene clusters as a result of the tandem duplication of an ancestral gene sequence. For example, the immunoglobulin genes are clustered at 14q32.33 (IGHA, IGHD, IGHG) and 22q11.2 (IGLC, IGLL), the T-cell receptor genes at 14q11.2 (TCRA, TCRD), 7q35 (TCRB) and 7p14-p15 (TCRG), the pregnancy-specific glycoprotein (PSG1, PSG2, PSG3, PSG4, PSG5, PSG6, PSG7, PSG8, PSG11, PSG12, PSG13) genes at chromosome 19q13.2, the histocompatibility antigen (HLA) genes at chromosome 6p21.3 whilst the three alkaline phosphatase genes (ALPP, ALPI, ALPPL2) are clustered at chromosome 2q37.

Some genes appear especially prone to duplicate, probably by virtue of their already being clustered in multiple copies. Thus, the carcinoembryonic antigen (PSG, CEA; 19q13.2) gene family has undergone multiple, but independent, multiplication events in both the rodent and primate lineages. In simiiar vein, a disproportionate fraction of mapped zinc finger gene family members are located on chromosome 19. Such clustering has probably arisen as a result of the serial duplication of a single ancestral gene on the same chromosome.

Members of gene families in different species may however be amplified differentially. For example, in the human genome, three members of the formylpeptide receptor gene family (FPR1, FPRL1, and FPRL2) cluster at chromosome 19q13.3, whereas in mouse, there are six *Fpr* gene on a syntenic region of chromosome 17. The human FPRL2 gene and four murine *Fpr* genes arose after the divergence of human and mouse.

Translocation of Duplicated Genes

Gene family members have often become chromosomally separated through translocation and it would appear as if the older the multi gene family, the more likely this is to happen. Thus, the chromosomally unlinked CD36L2 (chromosome 4), CD36L1 (chromosome 12) and CD36 (7q11.2) membrane glycoprotein genes were duplicated and diverged from an ancestral gene prior to the separation of the arthropod and vertebrate lineages. Other examples of ancient duplicated genes that have become separated by translocation are the thrombospondin

(THBS1, 15q15 and THBS2, 6q27) genes which arose ~900 Myrs ago and the transforming growth factor-β (TGFB1, 19q13.2; TGFB2, 1q41; TGFB3, 14q24) genes which arose ~30 Myrs ago.

Duplication of Translocated Genes

Genomic duplication may be followed by further local tandem duplication as exemplified by the human homeobox gene cluster 7p14-p15 (HOXA), 17q21-q22 (HOXB), 12q12-q13 (HOXC) and 2q31 (HOXD). The same phenomenon is exhibited by the human sodium channel genes many of which reside in the same paralogous chromosome segments as the HOX gene clusters: SCN1A, SCN2A, SCN3A, SCN6A, SCN7A and SCN9A (2q23-q24), SCN8A (12q13), SCN4A (17q23-q25), SCN5A and SCN10A (3p21-p24).

By contrast the localized duplication of translocated genes appears to be a general features of the human olfactory receptor (OLFR) genes. Thus, several OLFR gene clusters on chromosome 11 (11p15, 11p13, 11q24) are more similar to each other than to OLFR genes on chromosome 17, implying that translocation was followed by regional tandem duplication. However, the OLFR gene cluster on chromosome 17p13.3 contains members that do not belong to the same subfamily, suggesting that it originated instead by duplication of an entire gene cluster followed by translocation of that cluster.

Syntenic Relationships and Gene Dispersal

Gene duplication sometimes creates gene families whose members are both syntenic and dispersed, for example the human purinoceptor gene family, (P2RY1, 3; P2RY2, 11q13.5-q14.1; P2RY4, Xq13; P2RY6, 11q13.5, P2RY17, chromosome 14). Seven human matrix metalloproteinase genes cluster at chromosome 11q22.3 (MMP1, MMP3, MMP7, MMP8, MMP10, MMP12, MMP13), but the other family members are dispersed between many other chromosomes (MMP2, 16q13; MMP9, 20q11.2-q13.1; MMP11, 22q11.2; MMP14, 14q11-q12; MMP15, 16q21; MMP16, 8q21; MMP19, 12q14). Other examples include the human fucosyltransferase (FUT1, FUT2, FUT3, FUT5, FUT6, 19p13.3; FUT4, 11q21; FUT7, 9q34; FUT8, 14q23) and annexin genes (ANX1, 9q11-q22; ANX2, 15q21-q22; ANX3, 4q13-q22; ANX4, 2p13; ANX5, 4q28-q32; ANX6, 5q32-q34; ANX7, 10q21.1-q21.2; ANX8, 10q11.2; ANX11, 10q21.1-q21.2; ANX13, 8q24.1-q24.2).

In some cases, evolutionary conservation of post-duplicational clustering may be important for the coordinate regulation of individual genes by common control elements. Possible examples of this include the spermatid-specific protamine (PRM1 and PRM2; 16p13.2) genes, the platelet membrane glycoprotein (ITGA2B, and ITGA3; 17q21-q22) genes, the albumin gene family (ALB, AFP, AFM, GC; 4q11-q13), the pregnancy-specific glycoprotein (PSG1, PSG2, PSG3, PSG 4, PSG5, PSG6, PSG7, PSG8, PSG11, PSG12, PSG13; 19q13.2) genes and the fibrinogen α, β, and γ (FGA, FGB and FGG; 4q21) genes.

Individual duplicated genes may still exhibit synteny simply because recombination has not yet separated them. Thus, the paralogous pulmonary surfactant protein D (SFTPD) gene at 10q22-23 lies in very close proximity to the pulmonary surfactant protein A (SFTPA1) gene. Similarly, the paralogous the interferon α/β receptor gene (IFNAR1) is closely linked to the cytokine receptor B4 (CRFB4) gene on chromosome 21q22.1. Synteny may however be retained for very long periods of evolutionary time. Indeed, the human proteasome α2 subunit (PSMA2) and TATA box-binding protein (TBP) genes are linked on chromosome 6q27 and their orthologues are also syntenic in *Drosophila melanogaster* and *Caenorhabditis elegans*. Similarly, conserved synteny is also apparent between the human genes PIM1, RXRB, PBX2, NOTCH4 and TNXA at 6p21 and their orthologues in *D. melanogaster* and *C. elegans*. The evolutionary conservation of close linkage over such long time periods is highly unusual and implies the existence of underlying functional reasons.

Functional Redundancy and Post-duplication Diversification

Evolution is opportunistic and gene duplication provides the opportunity for structural and functional diversification. This is well exemplified by the origin of the α-lactalbumin gene (LALBA; 12q13). α-Lactalbumin may be regarded as essentially a mammalian 'invention'. It is a regulatory subunit of the enzyme lactose synthetase. Phylogenetic analysis has indicated that α-lactalbumin evolved from lysozyme (LYZ; chromosome 12) (a bacteriolytic enzyme present in both vertebrates and invertebrates) through a gene duplication event that occurred before the divergence of mammals and birds, and prior to the evolution of the mammary gland. The origin of the LALBA gene thus appears to have long preceded the acquisition of its modern function. Its rapid evolution in the mammalian lineage need not however have been solely due to its acquisition of a new function. Once its lysozyme activity was lost, many of the amino acids essential for the protein's original function were freed from selective pressure, allowing rapid change, not all of it necessarily adaptive.

The protein products of a gene duplication may sometime be targeted to different cellular locations. The human mitochondrial HMG

CoA synthetase (HMGCS2; 1p12-1p13) gene encodes an enzyme which catalyses the first reaction of ketogenesis whilst the cytoplasmic form of the enzyme, encoded by the HMGCS1 gene (5p13), catalyses the second step in cholesterol biosynthesis from acetyl CoA. The two genes are thought to have arisen from a gene duplication ~500 Myrs ago. The emergence of the two enzymes as distinct entities served to link the pathways of β-oxidation and leucine catabolism and created the HMG CoA pathway of ketogenesis thereby providing a lipid-derived energy source.

Gene duplication has also potentiated the emergence of *isozymes*, enzymes that catalyze the same biochemical reaction but which may differ from each other in terms of tissue specificity, developmental regulation or biochemical properties. Thus, in vertebrates, the two subunits of lactate dehydrogenase are encoded by two genes (LDHA; 11p14-p15 and LDHB; 12p12) and these subunits can be combined in such a way as to form five tetrameric isozymes (A_4, A_3B, A_2B_2, AB_3, and B_4) each with its own distinctive properties.

Iwabe et al. (1996) examined gene duplications during organismal evolution and concluded that most gene duplications giving rise to entirely novel functions predated the divergence of the vertebrate and arthropod lineages. Genes which encode protein that are localized to cell compartments (compartmentalized isoforms) emerged by duplications which predated the separation of animals and fungi. By contrast, genes encoding products with virtually identical functions but differing tissue distribution (tissue-specific isoforms) have undergone duplications independently in vertebrates and arthropods after divergence of the vertebrate and arthropod lineages. Iwabe et al. (1996) concluded that there was a good correspondence between molecular evolution at the level of the gene, and tissue and organismal evolution.

Several gene duplications have occurred during primate evolution. For example, the tandem duplication responsible for the creation of the γl-(HBG1) and γ2-(HBG2) globin genes occurred prior to the divergence of Old World from New World monkeys. The 5.5 kb duplicated segment is bounded by two related LINE elements suggesting that the duplication occurred via homologous unequal recombination. Perhaps it was the functional redundancy initially introduced by the duplication which created an opportunity for the γ-globin genes to evolve a fetal function to replace their original embryonic function. Post-duplicational redundancy is sometimes still apparent in some systems. For instance, the MyoD family of transcription factors involved in myogenesis in skeletal muscle still exhibits functional redundancy (between the proteins encoded by the MYOD1 (11p15.1), MYF5 (chromosome 12) and MYF6 (chromosome 12) genes) and this is probably indicative of overlapping functions.

The primate glycophorin genes (GYPA, GYPB, GYPE; 4q28-q31) are thought to have emerged by duplication, mediated perhaps by recombination between *Alu* sequences. GYPA probably represents the ancestral gene and is present in all primates studied. The GYPB gene is present in human, chimpanzee and gorilla but not orangutan and gibbon, whereas the GYPE gene is present in human, chimpanzee but intriguingly only 7/16 gorillas tested. The complement C4 (C4A, C4B; 6p21.3) genes and the cytochrome CYP21 (6p21.3) gene also emerged in the primate lineage as a result of a single duplication occurring prior to the divergence of the apes from the Old World monkeys.

There are several other examples of gene duplications occurring in only one mammalian order. For example the bovine-specific coglutinin (*Cgn1*) gene is homologous to the human pulmonary surfactant protein D (SFTPD; 10q22-q23) gene and is thought to have evolved by gene duplication in the Bovidae after their divergence from the other mammals. Sometimes the number and distribution of *Alu* sequences may help the reconstruction of the phylogeny of a gene duplication event as in the case of the duplication of the apolipoprotein CI (APOC1; 19q13.2) gene, timed at about 39 Myrs ago. Some partially homologous genes may have undergone a process of duplication and divergence but one or other copy may have either acquired or lost DNA sequence at some stage thereby limiting the extent of observable homology between them. One example of this is the human α-N-acetylgalactosaminidase (NAGA; 22qter) and α-galactosidase A (GLA; Xq) genes. Six of the seven GLA exons were identically positioned in the NAGA gene but there was no similarity between the predicted amino acid sequences of GLA exon 7 and NAGA exons 8 and 9.

Truncated Gene Copies

In some cases, only a portion of a gene may be involved in these intra-chromosomal duplication event [e.g. the polycystic kidney disease 1 (PKD1; 16p13.3) gene)]. This often results, as in the PKD1 example cited, in the generation of a linked pseudogene representing a partial copy of the parental source gene . However, truncated copies need not always be inactive. Thus, the melanin-concentrating hormone (PMCH; 12q23-q24) gene has become partially duplicated during primate evolution to generate two truncated copies (PMCHL1 and PMCHL2) which have been translocated to chromosome 5p14 and 5q12-q13 respectively. These variant gene copies possess open reading frames, are expressed in a distinctive tissue-specific fashion and, in the authors' opinion, may represent 'genes in search of a function'.

Duplication Polymorphisms

Some gene duplications occur as polymorphic variants in the human population, for example the red (RCP; Xq28) and green (GCP; Xq28) visual pigment gene. Human males with trichromatic vision typically possess one RCP gene, one, two or more GCP genes and an RCP/GCP hybrid gene. However, some males with apparently normal colour vision can possess 4 RCP genes and 6 or 7 GCP genes. This notwithstanding, only one GCP gene is normally expressed, probably as a result of the activity of a locus control region upstream of the RCP gene.

Other examples of duplicational polymorphisms include the α1-globin, (HBA1; 16p13.3;), ζ-globin (HBZ, 16p13.3;) Gγ-globin (HBG2; 11p15), haptoglobin (HP; 16q22.1) and proline-rich protein (PRB1, PRB2, PRB3, PRB4; 12p13.2) genes. Duplicational polymorphisms may even manifest as gene cluster copy number variation, for example that involving the CYP21, CYP21P, C4A, and C4B genes at human chromosome 6p21.3. Another example of a duplicational polymorphism is that of three members (one potentially functional) of the olfactory receptor (OLFR) gene family which occur in tandem within a block that is duplicated at 14 different subtelomeric location in the human genome. This results in normal individuals possessing between 7 and 11 copies of this block in their genomes. Trask et al. (1998a) suggested that sub-telomeric regions could serve as 'nurseries' for the generation of diversity by promoting gene duplication.

Other gross duplicational polymorphisms are found in the immunoglobulin V_H gene cluster (IGHV; 14q32): one of 50 kb in length is present in 73% of individuals and results in the gain of 5 functional V_H segments whilst another of length ~80 kb is present ~50% of individuals and involves the gain/loss of two functional V_H segments. A third such polymorphism involving only a single additional V_H segment has been reported to occur in 27% of individuals. A duplicational polymorphism is also apparent in the C_H gene cluster: the IGHG4 gene (14q32) is duplicated in 44% of haplotypes. Finally, the partial duplication of the GABA A receptor α5 (GABRA5) gene is polymorphic in that individuals differ with respect to gene copy number.

Intragenic Gene Duplications in Evolution

Numerous examples of human genes have now come to light in which the encoded proteins have emerged through the introduction of individual exons or blocks of exons. Some of these exons encode specific protein domains (e.g. zinc finger, homeobox, immunoglobulin-like, epidermal growth factor-like, fibronectin, ABC cassette, Sushi, ankyrin, chymotrypsin etc.) which have come to be distributed between a large number of different protein through *exon shuffling*. Protein evolution may however also involve the internal duplication or amplification of individual eons, blocks of exons or alternatively repetitive sequence motifs within exons. Some typical examples of such intragenic duplication event are explored below.

Multi-exon Duplications

Some genes encode proteins that comprise two homologous domains and are therefore likely to have originated through a gross internal duplication. Thus, the 17 exon human transferrin (TF; 3q21) gene is thought to have originated by an internal duplication which may have resulted from an unequal crossing over event. Similarly, the human angiotensin I converting enzyme is encoded by a gene (DCP1; 17q23) which comprises 26 exons that encode two homologous domains each containing an active site. Exon number and size in the DCP1 gene are consistent with an ancient internal duplication, as are the codon phases at the exon-intron boundaries. An ancient internal duplication probably also occurred in the cystic fibrosis transmembrane conductance regulator (CFTR; 7q31.3) gene which encodes a protein with two transmembrane domains and two nucleotide binding fold domains. By contrast, successive duplications of ancestral domains appear to have occurred

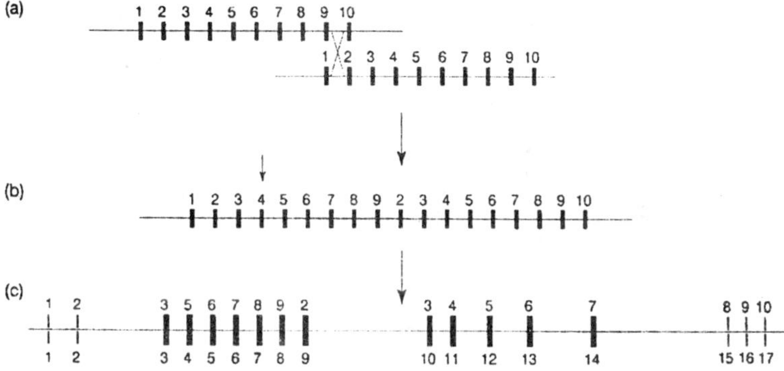

Fig. 13.9. A possible scheme for the evolution of the human transferrin (TF) gene.

in the human kininogen (KNG; 3q27) gene. Other examples of human proteins displaying internal domain duplication include calbindin (six 43 amino acid repeats), fibronectin (twelve 40 amino acid repeats), plasminogen (five 79 amino acid repeats) and α-tropomyosin (seven 42 amino acid repeats). Many other genes are also likely to have evolved by internal gene duplication but the duplicated regions have probably diverged so much over evolutionary time that sequence homology between them is no longer discernible.

Exon Duplication

It has been estimated that at least 6% of exons in human genes have arisen by the duplication of pre-existing exons. One example is that of the human CHC1 (1p36.1) gene, which encodes a protein involved in the coupling between DNA replication and mitosis. This gene comprises 14 exons, eight of which encode the 7 tandemly repeated domains of ~60 amino acids within the CHC1 protein; each repeat is encoded by a single exon except for repeat IV which is encoded by exons 10 and 11 separated by an inserted intron. The CHC1 gene

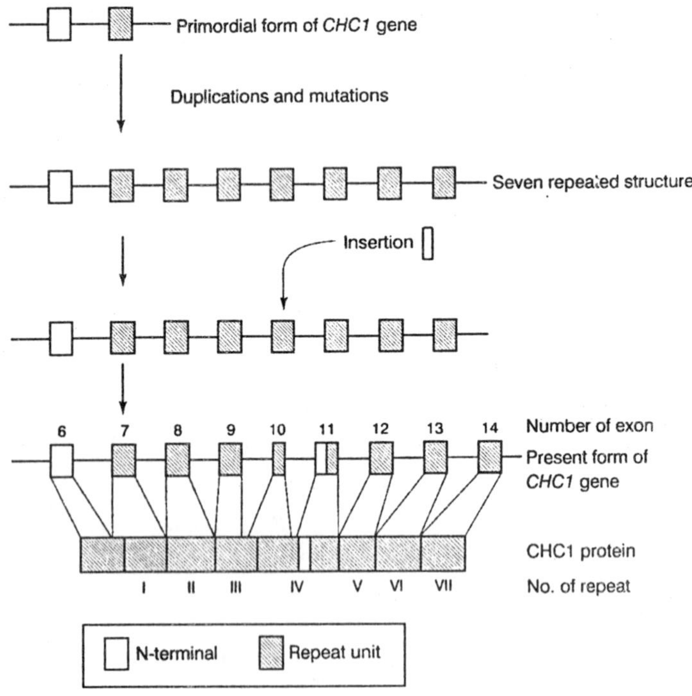

Fig. 13.10. Structure and evolution of the human CHC1 gene.

therefore appears to have arisen through the amplification of a primordial exon. Gene construction by exon amplification has also been employed in the macrophage mannose receptor (MRC1; 10p13) gene; 26 of the 30 exons of the MRC1 gene serve to encode the eight C-type carbohydrate recognition domains. Ceruloplasmin (CP; 3q23-q25) consists of three homologous repeat units and appears to have evolved by successive duplication of a primordial copper-binding protein domain of ~350 amino acid residues. Finally, the human annexin II (ANX2; 15q21-q22) gene encodes a protein containing 8 copies of a conserved 70 amino acid repeat.

Exon duplication/ amplification is clearly a common mechanism of evolutionary change and its occurrence may often be inferred from studies of exon/intron distribution or from the repetitivity of protein domains. Thus, the six paralogous genes of the human salivary proline-rich protein family closely linked on chromosome 12p13.2, differ from each other in terms of the number of copies of a 63 bp tandem repeat in exons 3 of the gene: PRH1 and PRH2 (6 repeats), PRB1 and PRB3 (15 repeats) PRB2 (16 repeats) and PRB4 (11 repeats). Examples of orthologous gene pairs in which only one orthologue manifests exon duplication/amplification are somewhat rarer. The oxidized low density lipoprotein receptor 1 (OLR1; 12p12.3-p13.1) gene serves to illustrate the principle; the rat *Olr1* gene encodes three 46 amino acid repeats between the transmembrane and lectin-like domains of the LOX-1 protein, whilst the human an bovine OLR1 genes encode only one such repeat.

The primate semenogelin (SEMG1, SEMG2) genes which, in human, are closely linked on 20q12-q13.1, encode the major protein constituents of the seminal fluid. These genes differ however between species as a result of the seminal fluid. These genes differ however between species as a result of internal duplications of ~180 bp segments encoding 60 amino acid repeats. Human semenogelin II contains two fewer repeats than rhesus monkey semenogelin II which appears to contain two fewer repeats than baboon semenogelin II. The primate semenogelin genes arose by duplication of an ancestral gene about 60 Myrs ago and the human semenogelin I gene appears to have lost two 60 amino acid repeats by comparison with the paralogous semenogelin II gene. The semenogelin I protein from cotton-top tamarin (*Saguinus oedipus*) possess three additional repeats as compared to human and, intriguingly, possesses 14 potential glycosylation sites not present in the human protein.

Most known genes manifesting internal duplication exhibit strong conservation of intron positions between the duplicated domains. One notable exception, however, is the rabbit phosphofructokinase gene. Sequence homologies between bacterial and rabbit phosphofructokinases, and between the amino and carboxy terminal ends of the rabbit enzyme are consistent with the origin of this gene being via a process of internal duplication and divergence. However, intron positions are not conserved even between the two halves of the rabbit gene.

Exon duplication may potentiate alternative splicing. Thus, the human ketohexokinase (KHK; 2p23) gene contains two very similar 135 bp exons (termed 3a and 3c) which are mutually exclusively spliced into KHK mRNA. Both the exon-intron structure and the pattern of alternative splicing are conserved between human, rat and mouse, consistent with the existence of two evolutionarily conserved KHK isoforms. This alternative splicing event is also tissue-specific since, in both rat and human, those tissues that express high levels of KHK incorporate exon 3c whereas other tissues incorporate exon 3a. Interestingly, a shift in splicing choice from exon 3a to 3c appears to occur during development between the human fetus and adult.

Intra-exonic Duplications

Internal gene duplications may sometimes be quite subtle. Thus, the 3' untranslated regions of the human and monkey cytochrome c oxidase subunit II (MTCO2; mitochondrial genome) genes have generated by duplication events involving a 14 bp region that occurred during primate evolution. The RNA derived from the MTCO2 gene has the potential to form stable stem-loop structures in a region immediately preceding the duplication site and these inverted repeat sequences may have played a role in promoting the duplicational events.

In some cases, homogenization of internal repeats has occurred as for example with the complement control protein repeats of the baboon and human complement receptor type 1 (CR1; 1q32) genes. This process has been termed *horizontal* or *concerted evolution* and the mechanism involved in likely to be either unequal crossing over or gene conversion. Whichever mechanism is responsible, the following example suggests that intragenic homogenization may be a less efficient process than intergenic homogenization. The human small proline-rich (SPRR) proteins, induced during keratinocyte differentiation, are encoded by two SPRR1 genes (A and B), approximately seven SPRR2 genes and a single SPRR3 gene, closely linked on chromosome 1q21-q22. The central segments of the encoded polypeptides are composed of tandemly repeated units of either eight or nine amino acids. Thus the consensus octamer PKVPEPCH is found 6 times in SPRR1A and SPRR1B, the nonamer PKCPEPCPP three times in SPRR2 and the octamer TKVPEPGC 14 times in SPRR3. This is consistent with a process of internal duplication that began after the divergence of the SPRR genes into three distinct subfamilies. It is evident that, during the evolution of the SPRR gene family, there has been a bias toward either intragenic or intergenic duplications. Thus in SPRR2, intergenic recombination has occurred more frequently than intragenic recombination since there are seven SPRR2 genes each with three repeats. By contrast, in SPRR3, no intergenic recombination has occurred but intragenic recombination has occurred frequently to generate 14 repeat copies. Interestingly the percentage of amino acid conservation (relative to the consensus for each type of gene) is significantly higher for the SPRR2 repeats than for either SPRR1 or SPRR3 suggesting that intergenic homogenization may be more effective than intragenic homogenization.

Concerted evolution is not however an obligatory property of internally repetitive proteins. Take for example the case of the α- and β-spectrins. Spectrin is a red blood cell cytoskeletal component which consists of a tetramer of two antiparallel ab spectrin dimers. The α-spectrin (SPTA1; 1q21) and β-spectrin (SPTB; 14q22-q23) genes evolved by duplication of a common ancestral gene which existed before the divergence of the vertebrate and arthropod lineages ~600 Myrs ago. The structure of both protein subunits is consistent with successive intragenic duplications. The α- and β-subunits consist of tandemly repeated segments of 106 amino acids of which 20 occur in α-spectrin, 17 in β-spectrin. Although the α-spectrin segments appear to have evolved in homogeneous fashion, β-spectrin segments exhibit considerable heterogeneity. One explanation for this difference is that some segments with specific functions may have evolved differently from others. Indeed, on the basis of the similar locations of the heterogeneous α- and β-spectrin segments, Muse et al. (1997) suggested that the α- and β-spectrins have co-evolved, and those segments that are intimately involved in subunit dimerization have been evolutionarily constrained by the structures of their binding partners. Muse et al. (1997) found no evidence for interrepeat exchanges and therefore concluded that neither gene conversion nor recombination had operated. At some stage, probably before the divergence of arthropods and vertebrates, concerted evolution may have operated but this initial phase probably ceased as the individual segments began to diverge at the DNA level.

Emergence of Primordial Genes by Oligomer Duplication

It is possible that primordial coding sequences emerged by a process of sequential duplication of base oligomers to yield 'genes' encoding polypeptides with significant periodicity. Ohno (1987) cites the example of a putative primordial rhodopsin gene, which gave rise to this ancient family of seven transmembrane domain-containing proteins including among others, bacterial rhodopsin and vertebrae retinal opsin, β_2-adrenergic receptor and muscarinic acetylcholine receptor. This primordial gene originally contained CCTGCTG, CCTGGCC and GCTGGCC heptameric repeats. Ohno (1987) showed that the gene encoding porcine muscarinic acetylcholine receptor still contains many of these oligomeric repeats. The original heptameric repeats appear to be more stringently conserved in those portions of the gene encoding the seven transmembrane domains whereas new repeat units comingle with old repeats in those portions that encode the extracellular and intra-cytoplasmic domains.

Intragenic Duplication Polymorphisms

Partial internal duplications of genes can also occur as polymorphic variants. A particularly dramatic example is provided by the human apolipoprotein(a) (LPA; 6q27) gene. This gene possesses at least 34 different alleles containing a variable number (between 12 and 51) of tandemly repeated kringle IV-encoding domains. Each allele gives rise to a different apolipoprotein(a) isoform thereby explaining the size polymorphism of the protein: 300-800 kDa. The LPA gene has been described as a 'plasminogen gene gone awry'; an apt description in that the closely linked and evolutionary related plasminogen (PLG; 6q27) gene encodes a protein with only 5 kringles. The LPA gene is thought to have arisen between 40 Myrs and 90 Myrs ago during the adaptive radiation of the mammals. At least ten kringle IV-encoding domains are present in rhesus macaque suggesting that kringle number may have expanded progressively during primate evolution. Although the LPA gene was originally thought to be confined exclusively to primates, it has also been found in hedgehogs in which it possesses 31 kringle-encoding domains. Since these are kringle III repeats, we must surmise that an apolipoprotein(a)-like gene arose independently from a plasminogen-like gene in this insectivore ($\sim$80 Myrs ago) and experienced a similar expansion of a subset of its kringle repeats to that found in humans—a remarkable example of *convergent evolution*. In humans, the LPA kringle repeat number polymorphism may not be without clinical significance because (i) there is an inverse relationship between apolipoprotein(a) isoform size and plasma apolipoprotein(a) levels and (ii) elevated levels of apolipoprotein(a) are associated with an increased risk of atherosclerosis and cardiovascular disease.

Several human mucin genes exhibit highly polymorphic tandemly repetitive regions, for example MUC1 (1q21), MUC2 (11p15) and MUC4 (3q29). Intragenic repeat copy number polymorphisms are also evident in the human proline-rich protein (PRH1; 12p13.2) gene where alleles vary in terms of the number of copies of a 63 bp repeat in exon 3 and the complement receptor type 1 (CR1; 1q32) gene where the two most common alleles differ in terms of the presence of a long homologous repeat of ~450 amino acids. Finally, in the human filaggrin (FLG; 1q21) gene, a length polymorphism of 10, 11, or 12 copies of a 972 bp repeat occurs within its polyprotein precursor which is subsequently cleaved into individual functional filaggrin molecules.

Coding Sequence Expansion and Contraction Resulting from the Introduction or Removal of Initiation and Termination codons

The mutation of initiation codons leading to the extension of the protein coding sequence of a gene is known to be a cause, albeit an infrequent one, of human genetic disease. Such mutations have however also occurred on a number of occasion during gene evolution. Thus, an ATG→GTC substitution occurred in the Met initiator codon of the ZNF80 (3q13.3) gene in the common ancestor of African green monkey and rhesus macaque which resulted in a change in the site of translational initiation to a location 20 codons amino terminal to the Met initiator codon used by humans, chimpanzees and gorillas. However, in African green monkey, the ZNF80 protein is only 213 amino acid residues in length (as compared to 273 residues in humans and the great apes and 293 residues in rhesus macaque) owing to truncation of the protein due to an additional GAG→TAG substitution introducing a novel termination codon.

Another example of mutation resulting in the specific-specific use of alternative initiation codons is provided by an ATG→ATA transition in the human hepatic peroxisomal L-alanine: glyoxylate aminotransferase 1 (AGXT; 2q36-q37) gene. This lesion removed the original initiation codon and an alternative downstream initiation codon is used in the translation of the human AGXT transcript. The rat *Agxt* orthologue encodes a protein which, by comparison with the human protein, possesses an extra 22 amino terminal amino acids specifying a leader sequence thought to contain a mitochondrial targeting signal that is absent in the human protein.

The mutational removal of a termination codon may also lead to the elongation of a protein product. One example of this is provided by the porcine P-glycoprotein genes whose coding regions may be seen to have been extended by such a mutation when compared to the human orthologous proteins (PGY1, PGY3; 7q21).

Minisatellites, Microsatellites and Telomeric Repeats

A detailed discussion of the evolution of the various families of satellite, minisatellite and microsatellite DNA would be somewhat tangential to the remit of this volume and would merely serve to recapitulate a previous volume in this series. However, a brief resume will be given to provide the necessary background to guide the interested reader to the relevant literature.

Minisatellite DNA Sequences

Minisatellite DNA sequences occur in all eukaryotes from yeast to human. Minisatellites frequently exhibit substantial allelic variability with respect to repeat number and allele length analysis has demonstrated germline mutation rates as high as 15% per gamete. Minisatellite mutation may involve intra-allelic rearrangements whose frequency, unlike inter-allelic rearrangements, is influenced by the size of the tandem array. Sequence similarities, manifested by a subset of minisatellites, to the Chi recombination promoting element of *E. coli* have led to the suggestion that this 'core sequence' might be recombinogenic and could serve to promoter unequal crossing over. However, analysis of flanking polymorphisms has not indicated the exchange of markers predicted for the products of unequal exchange between alleles. This notwithstanding, recombination hotspots sometimes co-localize with minisatellites which has led to the suggestion that minisatellite instability may be a by-product of meiotic recombination.

Monckton et al. (1994) have shown that minisatellite mutation can involve complex inter-allelic gene conversion events. These may exhibit polarity since the gain of a few repeat units appears to be confined to one end of the tandem repeat array. One alternative proposal is that minisatellite mutation may involve an array homogenization process which could operate by biased repair of intra-helical (slippage) or inter-helical (unequal sister chromatid exchange) heteroduplexes. Whatever the mechanism(s) underlining their allelic variability, the rapid evolution of minisatellites appears to have rendered them an important substrate for the opportunistic processes of molecular evolution. Indeed, several are known to have become recruited as gene regulatory elements.

Microsatellite DNA Sequences

Microsatellites typically mutate with a frequency of 10^{-3} to 10^{-4} although mutation rates vary between loci by several orders of magnitude. Microsatellite mutation events in the male germline outnumber those in the female germline 5- to 6-fold. Interestingly, alternative alleles at the same locus may differ dramatically in terms of their mutation rate. By contrast, microsatellite location is often conserved at orthologous positions in the genomes of different species compared allele length distributions for a considerable number of human microsatellites with their orthologues in chimpanzees, gorillas orangutans, baboons, and macaques and claimed a tendency for the loci to be longer in humans. Although some microsatellites are more variable in nonhuman primates than in humans, a tendency for the loci to be longer in humans would be consistent with the directionality of microsatellite evolution and compatible with evolution proceeding at different rates in different species. However, it could also be due to ascertainment bias in that it is precisely the longest, most mutable and therefore most informative of human microsatellites that have been selected as genetic markers. In principle, the large size of the human population could have compounded this effect since it could support a higher level of genetic diversity; in smaller populations, much variability is lost as alleles go to fixation. In practice, however, it would appear as if ascertainment bias cannot be the sole explanation for inter-specific differences in microsatellite repeat length. Thus, the tendency for microsatellite loci to be longer in humans may not simply be an experimental artefact.

Telomeric and Centromeric Repetitive DNA

Telomeric TTAGGG repeat number varies not only between human chromosomal arms but also between individuals. Indeed, there is some evidence in humans for the exchange of telomeric and subtelomeric repeats between nonhomologous chromosomal ends. Since satellite DNA sequences at the telomeric junctions of chimpanzees do not show any similarity to their counterparts in human or orangutan, it is likely that telomeres became reorganized relatively recently during primate evolution.

Alphoid satellite DNA is found as a tandem repeat in long chromosome-specific arrays at the centromers of all primate chromosomes. These arrays are highly variable within and between homologous chromosomes in the same species.

Expansion of Unstable Repeat Sequences

DNA sequences that are internally repetitive are particularly prone to misalignment during DNA synthesis. If such misalignment takes place, the nascent strand can slip back in multiples of the repeat unit and the newly synthesized DNA strand will be elongated by comparison with the parental strand. Since many coding sequences contain simple sequence repeats (for example, the genes encoding the 28S and 18S ribosomal RNAs (RNR1-5) and the TATA-binding protein TBP (TBP)), repeat expansion may have been involved in the evolution of these sequence . However, these simple repeats also have the potential to be involved in disease pathogenesis and it is from studies of genetic disease that most of our knowledge of triplet repeat expansion is derived.

Triplet Repeat Expansion Disorders

One manifestation of DNA slippage involving simple repetitive sequences is the instability of certain trinucleotide repeat sequences. This mutational mechanism underlies fragile X mental retardation syndrome, a condition associated with the presence of a fragile site on the X chromosome (FRAXA). The brain-expressed FMR1 (Xq27.3) gene responsible contains an unusual (CGG)n repeat in its 5' untranslated region. This repeat exhibits copy number variation of between 6 and 54 in normal healthy controls, between 52 and >200 in phenotypically normal transmitting males (the 'premutation') and between 300 and >1000 in affected males (the 'full mutation'). Thus a continuum exists between a copy number polymorphism present in the general population, the asymptomatic premutation which involves limited expansion of (CGG)n copy number, and the full mutation which appears to require copy number expansion beyond a certain threshold value. Expansion of premutations to full mutations is thought to be a prezygotic event and occurs only during female meiotic transmission. Alleles with a repeat copy number of <46 do not exhibit elevated meiotic instability. By contrast, for alleles with 52-113 repeat copies, the premutation expands to the full mutation in 70% of transmissions whereas the corresponding figure for alleles with >90 repeat copies is 100%. The probability of repeat expansion thus correlates with the repeat copy number in the premutation allele, consistent with a mechanism of slipped mispairing during replication. Expansion of a sequence can thus itself lead to further expansion, a process termed 'dynamic mutation' by Richareds et al. (1992). In FRAXA, triplet repeat expansion is thought to exert its pathological effects by down-regulation of FMR1 gene expression through hypermethylation of the promoter region upstream of the CGG repeat and repression of translation of the FMR1 transcript.

The discovery of this novel mutational mechanism has led to the recognition that the expansion of unstable triplet repeats is also responsible for a number of other human inherited diseases. These often manifest a wide range of clinical severity and possess unusual features such as increasing severity and penetrance in successive generations ('anticipation') and a sex bias in the transmission of the disease which correlates with the degree of meiotic instability and allelic expansion. The nature and location of the repeat sequence involved varies between disease states as does the extent of the expansion necessary to bring about symptoms of disease, and of course the mechanism of pathogenesis consequent to repeat expansion. It should however be noted that repeat expansion is not confined to triplet repeats; minisatellite expansion can also occur as in progressive myoclonus epilepsy and the clinically asymptomatic fragile sites FRA16A and FRA16B.

Triplet repeat expansions have so far been found to be associated mainly with three types of sequence: CAG (with complement CTG), CGG (with complement CCG) and GAA (with complement TTC). Sequence specificity implies a role for DNA secondary structure in the expansion mechanism. Such repeats are known to form stable hairpin loop structures which become more stable with increasing repeat number. DNA polymerase progression appears to be blocked by CTG and CGG repeats and the resultant idling of the polymerase may serve to catalyze slippage leading to repeat expansion.

A number of diseases are characterized by CAG repeat expansion within the gene coding region. This is thought to constitute a gain-of-function mutation through the incorporation of a polyglutamine tract into the protein product. Several factors are known to influence the stability of triplet repeats viz. the type of sequence, length of repeat, whether the repeat is interrupted or not, and the orientation of the repeat relative to the origin of replication. The removal of interrupting point mutations from repeat arrays is thought to be very important in promoting triplet repeat expansion and is significant in an evolutionary context as well as in cases of disease. Removal of these point mutations could occur simply by single base-pair substitution or by gene conversion or unequal crossing over. Other inherited conditions result from expansion of a triplet repeat in the 3' untranslated region (myotonic dystrophy; DMPK) and an intron and the resulting mechanisms of disease are consequently different.

Nature and Distribution of Triplet Repeats in the Human Genome

Trinucleotide repeats are 10 to 100-fold less frequent than (AC)n repeats and different types of trinucleotide repeat occur at different frequencies. Thus repeats (AAT)n and (AAC)n are the most frequent trinucleotide repeats found in the human genome, with (AAT)n exhibiting a high degree of copy number polymorphism. Both (CAG)n and (CCG)n repeats appear to be over-represented in the human genome, the former being polymorphic in number in the genomes of humans and non-human primates. (CAG)n repeats are rare in intronic regions, possibly because of their similarity to the acceptor splice site consensus, CAGG. Long homopeptides are present in 1.7% of human protein-coding sequences.

A number of human genes contain polymorphic triplet repeats (e.g. cadherin 2, (CDH; 18q12), breakpoint cluster region (BCR; 22q11), glutathione-S-transferase (GSTA1; 6p12), Na^+/K^+ ATPase β1-subunit (ATP1B1; 1q22-q25) but these repeats are not known to exert any pathological effect. Other genes have been identified solely on account of their possession of polymorphic trinucleotide repeats and these genes represent candidate loci for involvement in complex diseases. Trinucleotide repeat containing genes are also found in other species, including mouse, but no nonhuman example of triplet repeat expansion as a cause of a genetic disease has yet been documented.

Replication slippage involving short GC-rich motifs ('expansion segments') has occurred during the evolution of the vertebrate genes encoding the 28S and 18S ribosomal RNAs. Interestingly, different segments of the 28S rRNA subunit gene appear to have coevolved by 'compensatory slippage' allowing RNA secondary structure to be conserved as a consequence of runs of sequence motifs in one region being compensated for by complementary motifs in another.

Origin of Expanded Triplet Repeats

In myotonic dystrophy, all Caucasian and Japanese DM chromosomes possess a specific haplotype, whilst a second disease-associated haplotype has been found in Africans. Similarly, haplotype analysis has pointed to a single origin for Japanese and Caucasian

Machado-Joseph disease chromosomes, a single origin for Japanese and Caucasian dentatorubral and pallidoluysian atrophy chromosomes, a single origin for the Friedreich ataxia expansion, at least two origins for the Huntington disease expanded repeat and a small number of FRAXA progenitor chromosomes.

One of the best examples of the evolutionary emergence of an expanded triplet repeat is that found in the SRP14 gene (15q22) encoding the 14 kDa *Alu* RNA-binding protein. The human protein is larger than that of mouse and dog on account of an extra 28 residue alanine-rich C-terminal tail which is translated from a 3' GCA-rich trinucleotide repeat. In the prosimian *Galago*, the relevant sequence at the site of the human triplet repeat is GCA GCA, whereas the mouse possesses the sequence CCA GCA. By contrast, the African green monkey possesses 33 GCA repeats, whilst the owl monkey possesses 52 suggesting that a CCA→GCA substitution occurred in an ancestral prosimian thereby creating two consecutive GCA codons which then facilitated GCA expansion in higher primates. Interestingly, however, no intra-specific variability in repeat size was detected in primates.

Evolution of Repeat Number in the Genes Underlying Disorders of Triplet Repeat Expansion

The normal range for CAG repeat number in the HD gene is 8-35. Although modal CAG repeat length is fairly similar between different human populations, the degree of spread varies, being greatest among Africans and lowest among the Japanese. The breadth of this normal range suggests that natural selection is acting weakly if at all on HD alleles below the disease threshold. However, the population distribution of CAG repeat number in the HD gene exhibits an apparent asymmetry in that more alleles lie above rather than below the modal length. Using computer simulations, Rubinsztein et al. (1994) have shown that the distribution of HD alleles in human populations is explicable in terms of a simple length-dependent mutational bias. The observed distribution of alleles is thus explicable merely in terms of mutation and genetic drift thereby obviating the need to invoke positive selection. It may be however, that coding sequences can tolerate CAG repeat-encoded polyglutamine tracts relatively well, thereby minimizing the effect of negative selection. More controversially, in the case of myotonic dystrophy and Machado-Joseph disease, *meiotic drive* (also termed *segregation distortion*), the excess recovery of one of a pair of alleles in the gametes of an heterozygous parent, has been proposed as being responsible for maintaining the frequency in the population of chromosomes bearing triplet repeats capable of expansion into the disease range.

Djjan et al. (1996) examined the disease-associated CAG repeats in the HD, MJD, AR, and SCA1 orthologues of various nonhuman primates species. For the HD and MJD genes, CAG copy number was polymorphic in the nonhuman primates but there was essentially no overlap with the normal human range, findings also reported by Limprasert et al. (1996). The AR gene was also found to be polymorphic in nonhuman primates but with minimal overlap with the normal human range. For the SCA1 gene, the average repeat copy numbers exhibited by *Pan*, *Gorilla*, *Hylobates*, *Macaca*, and *Cercopithecus* were within the normal human range but were lower than the modal number is lower in nonhuman primates than in humans. From what we know of the propensity of CAG repeats to expand, this is much more likely to be due to a higher rate of expansion in humans than the alternative: contraction in the other primate species. In the case of the HD gene, this interpretation is supported by the finding of a similar number of CAG repeats in the murine HD gene to that found in the nonhuman primate HD genes. The high number of CAG repeats in the human HD gene is therefore due primarily to expansion within the human lineage. This contrasts with the expansion of CAG number in the AR gene which began earlier in hominoid evolution; the great apes contain an expanded CAG repeat not found in rodents. Thus, the AR CAG repeat expansion began prior to ape-human divergence but continued in human after divergence from the great apes. The most dramatic example of CAG expansion is however that of the involucrin (IVL; 1q21) gene which is so rich in CAG repeats and codons derived from CAG that it is likely that it is descended from a simple poly (CAG) sequence.

The CGG repeat in the 5' UTR of the FMR1 gene has been studied in 44 mammalian species form 8 different orders. The presence of this repeat in all species examined indicates that the CGG repeat has been conserved for over 150 Myrs and is therefore likely to be have some functional significance. Repeat length was found to be similar among the 24 nonprimate species, ranging from 4 to 12 units (mean 8.01 $\pm$ 0.8). By the contrast, the mean length of the repeat among the 20 primate species examined was 20.1 $\pm$ 2.3. Copy number polymorphism was not found to be limited to human, with *Ornythorhyncus* (platypus), *Artibeus* (phyllostomid bat) and *Pan* (chimpanzee) all possessing polymorphic repeats. Parsimony analysis predicted that the early mammalian CGG repeat was short (4-9 units) and uninterrupted, whereas an increase in copy number beyond ~20 repeats appears to have occurred at least three times independently in the Catarrhini. These expansions in the hylobatid apes, great apes and the cercopithecoid monkeys were associated with the addition of specific interspersions, CGA, AGG and CGGG respectively. These interspersions are unlikely to have arisen through DNA polymerases slippage and may have been mediated by unequal crossing over or gene conversion.

A comparison of trinucleotide repeats in human versus rodent coding sequences has indicated that, by and large, orthologous repeats have not been conserved for long periods of evolutionary time, either in terms of their size or location. As yet, there are no known pathological equivalents of human dynamic mutations in other species but this may merely reflect bias of ascertainment. Thus, nonhuman examples of disease-associated triplet repeat expansions may not yet have come to our attention and it may be that these expansions have no counterpart in the human orthologues. Similarly, the observation that triplet repeats associated with human disease are sometimes found to be both polymorphic and expanded beyond the mammalian ancestral state in non-human primates may also be viewed in terms of a bias of ascertainment rather than any intrinsic propensity of triplet repeats to expand in primate genomes. Finally, noting that most of the expanded triplet repeats are found in genes expressed in the nervous system, Hancock (1996) proposed that triplet repeat expansion may have occurred in parallel with the increase in brain complexity so characteristic of the human lineage and may even have been instrumental in bringing about this increase in complexity.

Unique Case of Involucrin

Involucrin is a unique protein in that it has evolved very rapidly in the primate lineage through changes in the length and copy number of tandem repeats within the coding region with the result that a full two thirds of its coding region has been created within the anthropoid lineage. The evolution of this gene/protein is therefore worth examining in some detail.

Involucrin is the most abundant protein component of the keratinocyte envelope and in cross-linked to membrane proteins by a transglutaminase. The coding region of the human involucrin (IVL; 1q21) gene lies within a single exon and encodes a 585 amino acid glutamine-rich protein. The human IVL gene is extremely rich in CAG repeats (encoding glutamine) and codons derived from it (CAG and CTG). Thus, CAG codons, and codons one substitution removed from

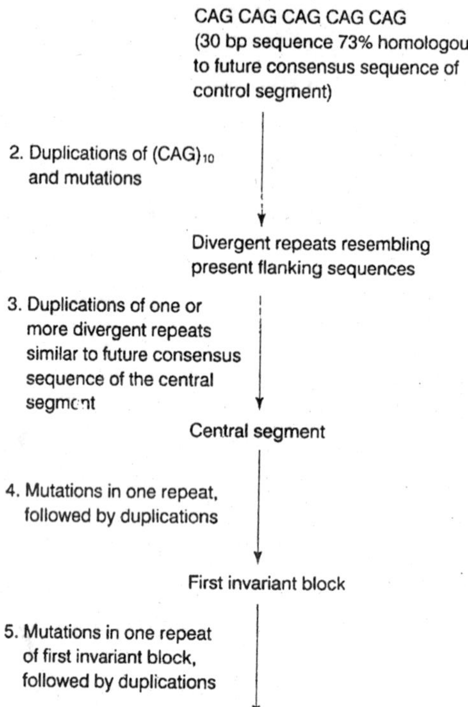

Fig. 13.11. Proposed scheme for the evolution of the human involucrin (IVL) gene.

CAG, comprise 64% of the total codon number. It is possible therefore that this gene is descended from a simple poly(CAG) sequence which has been subsequently modified by nucleotide substitution.

The time of origin of the involucrin gene is unclear but the gene is present in a wide variety of mammals and may be present in all terrestrial vertebrates. The length of the involucrin molecule has continued to grow by successive addition of repeats from prosimians (average 409 residues) through New World monkeys (average 488 residues) and Old World monkeys (average 514 residues) to hominoids (average 632 residues). The coding region outside the segment of repeats has by contrast changed very little in length. Since glutamine residues serve as amine acceptors in the transglutaminase-catalyzed cross-lining, the primordial glutamine-rich involucrin may have been able to function as a substrate for transglutaminase.

Involucrin gene sequences have now been determined in a range of primate and nonprimate species. Two different sites have been found within the coding region that contain variable numbers of repeat segments. One, at site P, lies ~80 codons from the ATG, contains a 16-amino acid repeat and is evident in nonprimate mammals, prosimians and tarsioids. The other, at site M, containing a 10-amino acid repeat, is located a further 68 codons C-terminal to site P and has increased in size in the anthropoid apes concomitant with a reduction in size of the P repeat segment.

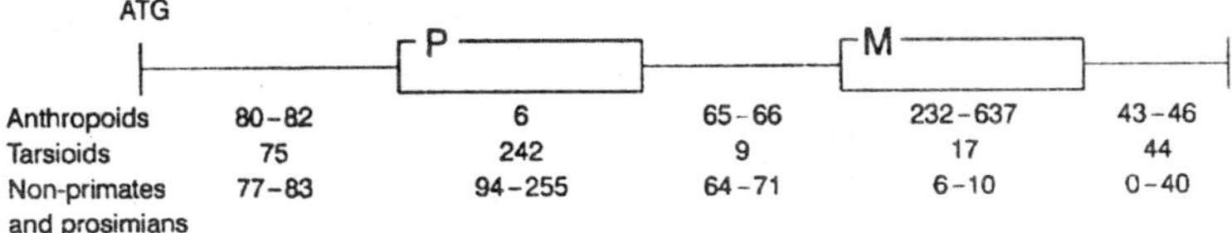

Fig. 13.12. Coding regions of primate involucrin genes illustrating the relative locations of repetitive regions P and M.

The involucrin genes of the dog, pig, and mouse have 6, 13, and 21 repeats, respectively at site P whilst the lemur (*Lemur catta*) and tarsier (*Tarsius bancanus*) possess 19 and 18, respectively. Within each species, however P repeat segments differ; thus three codons have been deleted in eight of the 13 repeat copies in *Galgo crassicaudatus* whilst two codons have been deleted in all repeat copies in *T. bancanus*. Consensus sequences also differ between nonanthropoid species in terms of nucleotide differences that occurred in any position in a codon. Changes in a consensus nucleotide must have arisen from the occurrence of a nucleotide substitution in one repeat followed by the correction of the analogous nucleotides at the corresponding positions in the other repeats, with adjacent repeats tending to be more homogeneous. Green and Djjan (1992) have proposed that a form of intragenic gene conversion may have been responsible for this phenomenon. Although the P segment was retained by prosimians and tarsioids, it was modified by site-specific deletions, the addition of certain repeats and nucleotide substitutions which served to alter the consensus codons of the repeats by a process of correction operating between neighbouring repeats. It is possible that the retention of the P segment was essential for the function of the protein since repeats at site M were lacking.

In the anthropoid apes, the M segment increased in size concomitant with the dramatic reduction in size of the P repeat segment. Perhaps deletion of the P segment was only possible after a sufficient number of repeats had been generated within the M segment.

Repeat Number	Type	1	2	3	4	5	6	7	8	9	10	Number of altered codons per repeat
10	A	K	H	L	E	Q	Q	E	G	Q	L	5
9	B	E	L	P	E	Q	Q	V	G	Q	P	10
8	A	K	H	L	E	Q	E	E	K	Q	L	10
7	B	E	L	P	E	Q	Q	E	G	Q	L	3
6	A	K	H	L	E	K	Q	E	A	Q	L	2
5	B	E	L	P	E	Q	Q	V	G	Q	P	3
4	A	K	H	L	E	Q	Q	E	K	Q	L	2
3	B	E	H	P	E	Q	Q	E	G	Q	L	7
2	A	K	H	L	E	Q	Q	E	G	Q	L	6
1	A	K	D	L	E	Q	Q	E	G	Q	L	10
												Mean 5.8

Fig. 13.13. Consensus amino acid sequences for each of the 10 repeats of the early region of the M segment of involucrin.

This notwithstanding the M segment continued to grow by successive addition of repeats from prosimians through New World Monkeys and Old World monkeys to hominoids. However, the repeats have the same consensus sequence in all the anthropoid apes. The M segment has been divided into early, middle and late regions which have been added vectorially in a 3'→5' direction and the repeats from these regions are shared to differing extends by different anthropoid species. Thus, the early region is common to all anthropoids, the middle region is shared by different species but to a lesser extent, and the late region is species-specific and must have developed after the divergence of the different species. After its divergence from the cercopithecoids, the hominoid lineage added 10 repeats to the middle region. After divergence from the common lineage, *Hylobates*, *Pongo*, and *Homo* acquired species-specific repeats in the late region. Another shared repeat segment is the early/middle (e/m) extension 5' to the late region which was probably formed independently in Old World and New World monkeys.

The site of repeat addition, which in the common anthropoid lineage was located in the P segment, moved in a 5' direction during the evolution of both the New World and Old World monkeys after their divergence. Deletions of repeats also took place within the early, middle and e/m regions of the M segment although these were relatively few in number as compared to the additions. These deletions have occurred independently of the vectorial process of repeat addition both spatially and temporally.

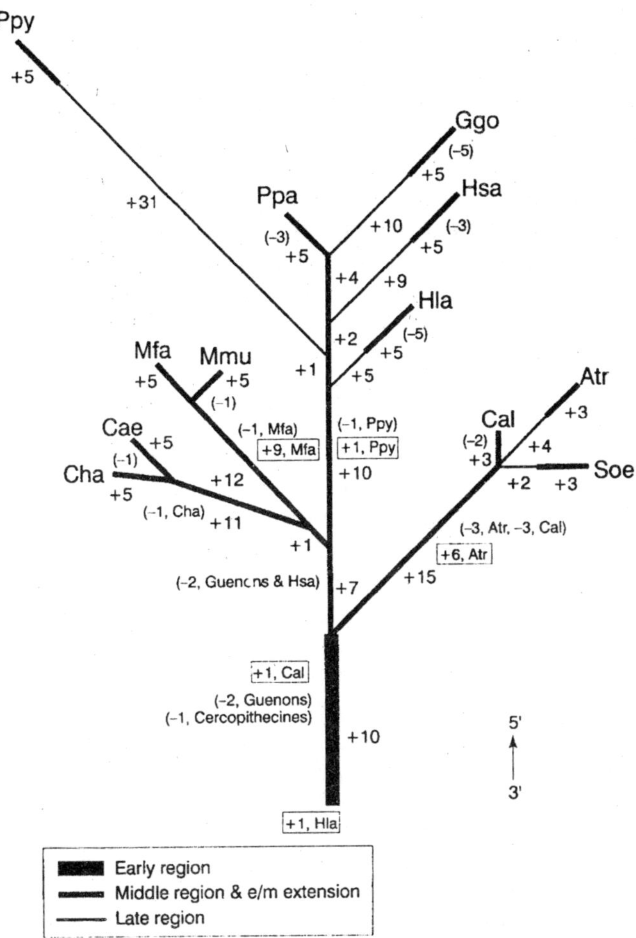

Fig. 13.14. Evolutionary tree of the M region of the primate involucrin gene detailing repeat additions.

In addition to the inter-specific differences in the size of the involucrin molecule, the higher primates also exhibit size polymorphisms which result from variation in the number of repeats in the late region. In humans, the polymorphism comprise variable numbers of B and B^S repeats within the late region. The most common allele in white Caucasians contains nine repeats ($3B^S$, 5B, 1A) but a second 'Mormon' allele ($2B^S$, 5B, 1A) has also been described. Blacks possess the 'Mormon' allele plus three other alleles containing 6, 7, or 9 B repeats. Repeat number polymorphisms in the late region have also been reported in *Aotus trivirgatus*, *Macacca mulatta* and *Gorilla gorilla*. The repeat pattern observed in anthropoid apes is explicable in terms of the presence in the involucrin gene of a hotspot at which repeats are generated by unequal recombination. In the human lineage, the location of this hotspot varies between racial groups such that in whites it is located within the B^S repeat region whilst in blacks, it is within the B repeat region. Thus, although the process of vectorial repeat addition has operated in both blacks and whites, there has been a difference between these racial groups in the sites of repeat addition, a finding that is consistent with an endogenously controlled mechanism.

Attempts have been made to relate the evolution of a larger involucrin molecule with a different repeat structure to the trend in anthropoid apes towards relative hairlessness. The P segment of repeats could conceivably have conferred some selective advantage during the early stages of mammalian and primate evolution as might the M segment when it emerged. At a biochemical level, natural selection might have favoured the retention of repeat-bearing segments in involucrin so as to provide a substrate for transglutaminase. Once the M segment has been generated, the repeats at site P could then have been deleted without selective disadvantage. However, there is a very considerable degree of latitude between different species in terms of what constitutes an effective transglutaminase substrate. In a comparison of the involucrins of 19 mammalian species, only 3.1% of amino acid residues were uniformly conserved. Thus, on balance, neither the very high rate of evolutionary change nor the observed patterns of mutational change in the involucrin gene provide convincing evidence for natural selection being the major motive force behind involucrin evolution in higher primates. Indeed, it is unlikely that natural selection could account on its own for either the highly variable pattern of repeat addition or the numbers of repeats added.

Natural selection and neutral mutation/genetic drift are both processes that govern how mutations, once they have arisen, spread through a population with the possibility of eventually becoming fixed. By contrast, the evolution of involucrin appears to be primarily mechanism-based, being dominated by endogenously controlled and spatially targeted mechanisms of mutagenesis. Such mechanisms are likely to be the major factor responsible for the *directed evolution* of involucrin, directed in the sense that the mechanisms promote change in a specific direction (i.e. increases in repeat number or gene conversion between homologous repeats) without the necessary assistance of natural selection.

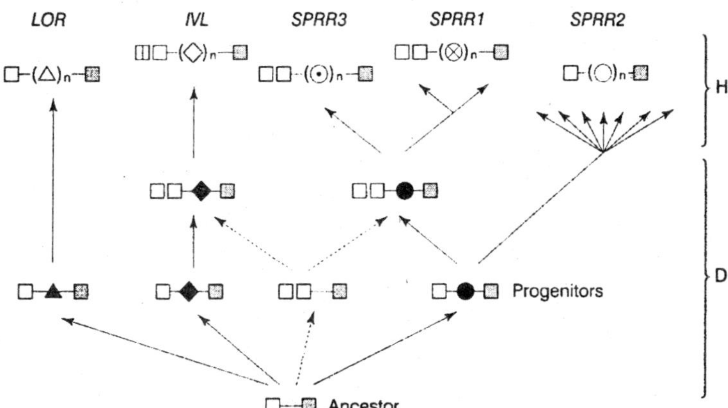

Fig. 13.15. Hypothetical scheme for the evolution of the involucrin (IVL), loricrin (LOR) and small proline-rich protein (SPRR) genes from a single ancestral gene.

Interestingly, several other human genes encoding epithelial proteins are located on the long arm of chromosome 1 [filaggrin (FLG), loricrin (LOR) epithelial mucin (MUC1), and the small proline-rich proteins 1A (SPRR1A), 1B (SPRR1B) 2A (SPRR2A) and 3 (SPRR3)]. These genes share a similar structure, contain multiple tandem repeats and are believed to be evolutionarily related both to involucrin and to each other. The LOR, FLG and MUC1 genes also exhibit repeat copy number polymorphism. The apparent association between chromosomal location and copy number polymorphism almost certainly reflects the possession by evolutionarily related proteins of a repeat structure that is particularly prone to the process of vectorial addition of repeats.

14

Evolutionary Molecular Events

The organization of genes into genomes is a large question, and this chapter looks at only two aspects of this issue. Both involve non-Mendelian inheritance, in which genes are copied laterally through the DNA rather than vertically from parent to offspring. We first consider concerted evolution by looking at the definition of a gene family, and how such families originate by duplication. We then see that gene families often show an evolutionary pattern that is difficult or impossible to explain if mutations are independent in each gene of the family; it can be explained if mutations move laterally, however. Next, we move on to the evolution of selfish DNA—that is, functionless sequences of DNA that are passively copied from generation to generation. We see that large quantities of non-coding DNA are present in eukaryotic species, and that (at least in some species) much of it is repetitive, being made up of duplications of certain unit sequences. We look at the kinds of repetitive DNA and the genetic processes that operate in them, such as slipping and unequal crossing over among tandem repeats, and transposition between scattered repeats. If non-coding repetitive DNA is selfish DNA, it would explain the apparently unnecessary excess of DNA in eukaryotes.

Evolution of Whole Genome

The theory of population genetics has been concerned with discrete genetic loci that are inherited according to Mendel's laws. The theory shows how mutation, selection, drift, migration, and linkage determine changes in gene frequencies. It was first developed in the 1920s and

1930s, before the dawn of molecular genetics. Since the 1960s, however, it has become increasingly clear that only a part—and probably a small part, at that—of the DNA in an organism consists of discrete genes, each coding for a protein or having some regulatory function. Large amounts of non-coding DNA are present as well, which are often arranged in the form of repeating unit sequences. The genes themselves, moreover, can have distinctive arrangements on the chromosomes; related genes, for instance, may be distributed in clusters in the DNA. These discoveries do not contradict the theory of population genetics. Indeed, the theory of population genetics is the only theory we have to make sense of them. However, additional concepts are needed to explain this system. In this chapter we will concentrate on two such concepts: concerted evolution and selfish DNA. They are related in that both depend on non-Mendelian heredity.

Any theory of how a genetic (or any other) phenomenon evolves must specify how the phenomenon originated, and then how it spread through the population. In the standard theory of evolution, new forms originate as mutations at a locus and then spread by either natural selection or random drift. The ideas in this chapter describe mechanisms for the origin of variants, rather than for their spread through a population. Once the new genetic variant of the types discussed in this chapter has originated, it spreads or does not spread by the same processes of drift and selection as any other genes.

Arrangement of Genes

How are the many genes of an organism arranged in its DNA? The theory of population genetics would make good sense if single genes were scattered arbitrarily. Some genes do exist by themselves as single copies, but many genes are arranged along the chromosomes in groups of related genes. The groups are called *gene clusters*. Related genes may be arranged in more than one physical cluster, and a whole family of related genes is called a *gene family*. Gene clusters and gene families may vary in importance in different taxonomic groups. They seem to be much rarer, for example, in insects than in mammals, so most of this chapter will be concerned with mammalian and vertebrate genes. Gene clusters, where they exist, have peculiar evolutionary properties, which we can introduce by means of two well-studied cases: the ribosomal RNA genes and the globin gene family.

Eukaryotic ribosomes are made up of three classes of ribosomal RNA: 5S, 18S and 28S. The 18S and 28S parts are formed from a large 45S precursor molecule that is transcribed from the DNA as a

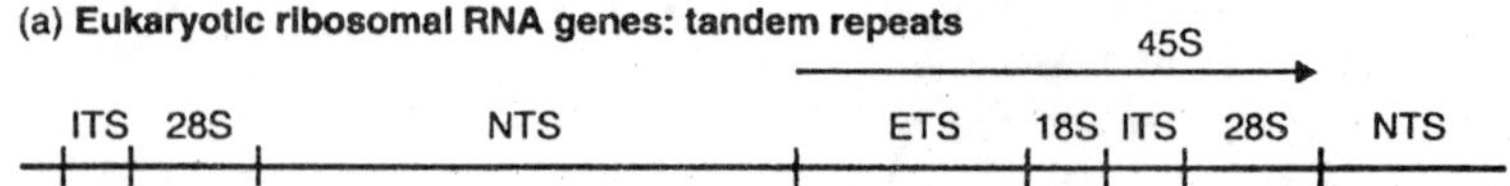

(b) Globin gene family: gene clusters

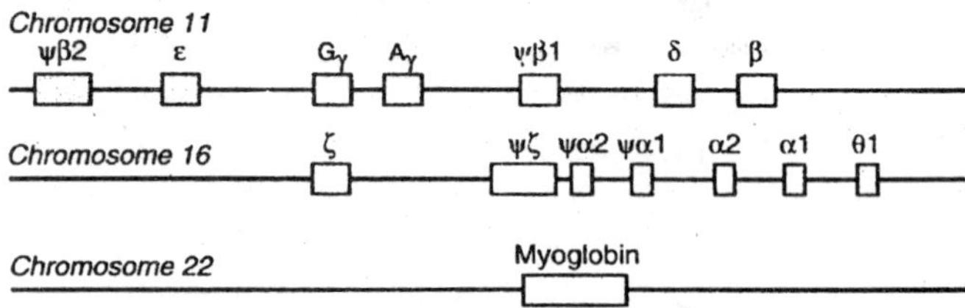

Fig. 14.1. Eukaryotic ribosomal RNA genes consist of long rows of tandem repeats of a unit genic organization. (b) The globin gene family consists of a number of clusters of related genes.

whole. Along the DNA, the 45S ribosomal genes are arranged in rows of multiple copies of a recognizable unit; hundreds of these units are strung out in a row of *tandem repeats*. In the African clawed toad *Xenopus laevis*, the 400-600 tandem repeats all appear on one chromosome; in humans, approximately 300 repeats are scattered through five chromosomes in tandem repetitive blocks of various sizes. The organism presumably needs multiple copies of the gene to synthesize large quantities of the gene product. Other functional genes also exist in tandem repeats. The five histone genes, at least in *Drosophila* and sea urchins, have this structure. Although the histone genes of vertebrates are also a multiple gene family, they are scattered about, rather than appearing side by side.

The globin gene family has a slightly different arrangement. The hemoglobin molecule that transports oxygen in the blood comprises a number of components. Human hemoglobin is a tetramer. In an adult, it is made up of two α-globins and two β-globins; in the fetus, of two α- and two γ-globins; and in the embryo, two ∈- and two ζ-globins. The sequences of these five globin types are similar and the genes that code for them occur in clusters on the chromosomes. In humans, two main clusters of globin genes exist: the α-globin cluster on chromosome 16 and the β-globin cluster on chromosome 11; the myoglobin gene on chromosome 22 is also part of the same family. Unlike the ribosomal RNA genes, the globins are not simple tandem repeats of an identical set of genes. Instead, the different globin genes represent distinct genes with distinct sequences. The ribosomal RNA genes and the globin genes, however, share the common property that the members of the gene family are linked in clusters. Many other examples of gene clusters are known, including the homeobox genes that are important in development and the HLA genes.

Gene Clusters

How could a cluster of genes, like the globin gene family, originate? For related genes strung out along a chromosome, duplication seems the most likely mechanism. Gene clusters are so common that duplication must have been a highly important mechanism during evolutionary history. Duplications probably arise by "mistakes" in recombination; the process is called *unequal crossing over*. A misalignment occurs, such that if two copies of a gene initially appeared on each strand, then after crossing over three copies appear on one

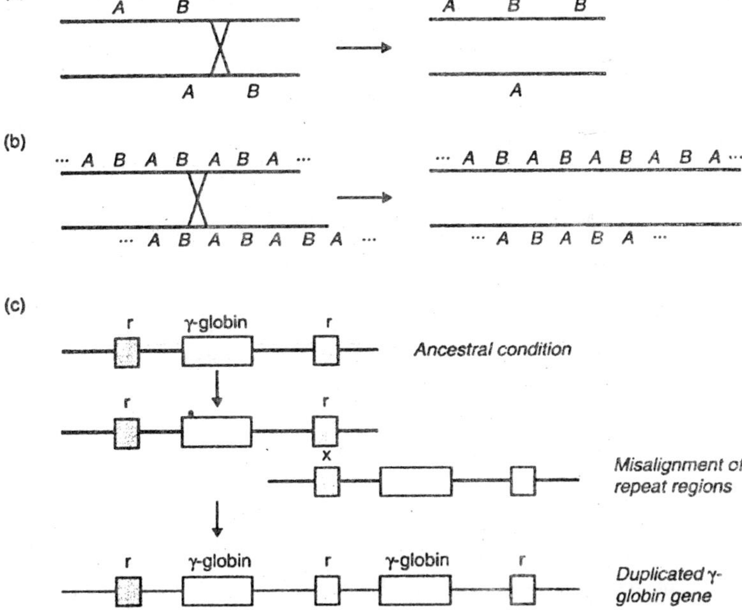

Fig. 14.2. Unequal crossing over occurs when the sequences of the two chromosomes are misaligned at recombination. (a) In a simple case, chromosomes with three copies and one copy of a gene could be generated from two chromosomes with two genes each. (b) In practice, misalignment is more likely if a long series of copies of similar sequences is present. (c) The human γ-globin genes include a repeat sequence (solid box) on either side of the gene itself (open box); this repeat could have been mismatched, causing duplication.

strand and one copy appears on the other strand. It is particularly easy to envisage a misalignment if a number of copies of a gene already occur in a row. The sequences of the two strands are matched up when they align for recombination and a gene could easily align with a non-complementary copy in the series of repeats.

In the human globin genes, duplication may have been facilitated by a non-coding repeat sequence between the globin genes. For example, the two γ-globin genes on chromosome 11 each have a short repeat sequence on either side. If this repeated sequence existed before the duplication, it could have been the target for the mismatching that produced the duplication. Certain variants of the human α-globin genes provide even more evidence for the importance of unequal crossing over among the globins. A normal human has two α-globin genes, but individuals can have three, or a single, α-globin gene. The obvious explanation is that these copies were generated by unequal crossing over.

Once a duplication has originated in a single copy, it could spread through the population by drift or selection. After the duplication became fixed, the two genes could retain the same sequence (like in ribosomal RNA genes) if selection favoured multiple copies of the gene. Alternatively, they could diverge to form a gene family (like the globins). A third possibility is that if only one of the copies of the gene is needed, the other might be switched off or decay by mutation and drift.

Heterozygous advantage at a locus may set up selection for gene duplication. Suppose that selection at a locus favours heterozygotes because an individual is better off with a two versions of a protein—perhaps because each protein is adapted to slightly different conditions. While these two versions are supplied by a single locus, disadvantageous homozygotes are generated when two heterozygotes breed together. If the gene were duplicated, the two versions could be supplied by homozygotes for each at the two loci, and the disadvantageous segregational load disappears.

The human globin gene family, as we saw, is arranged in two (or three, including myoglobin) clusters on two (or three) chromosomes. The kind of duplication may well explain the origin of many genes within each cluster, but is less likely to explain how more than one such cluster originates. Multiple clusters are more likely to have originated by translocation, by the duplication of part or all of a chromosome, or by polyploidy (i.e., the doubling, or multiplying, of the whole genome). For instance, suppose that one pair of chromosomes did not segregate equally at meiosis. One of the daughter cells would acquire a double set of the chromosome and the other would acquire none. The latter offspring would probably die, because it is missing the genes of the chromosome. Thus, the result is ultimately a mutation to a double set of the chromosome. As with gene duplication, after a chromosome has duplicated, its new set of genes could be maintained, diverge, or be suppressed, according to the outcome of drift or selective circumstances.

The globin gene family in two species of *Xenopus* supports this interpretation. In *Xenopus tropicalis*, unlike mammals, the α- and β-globin clusters are on the same chromosome. *X. laevis* (a tetraploid species), however, has two sets of the linked α- and β-globin gene cluster. If the α set were suppressed on one chromosome and the β set on the other, the mammalian arrangement would have evolved.

The sequences of the human globin genes can be arranged in a phylogeny. The phylogeny suggests that at least seven duplications have occurred in the history of the molecule. Once we have the tree, it is easy to use other molecular clock evidence to calibrate this scheme and estimate the times of the duplications. The difference in the sequence of the α- and β-globins, for instance, suggests they are derived from a duplication that happened approximately 500 million years ago.

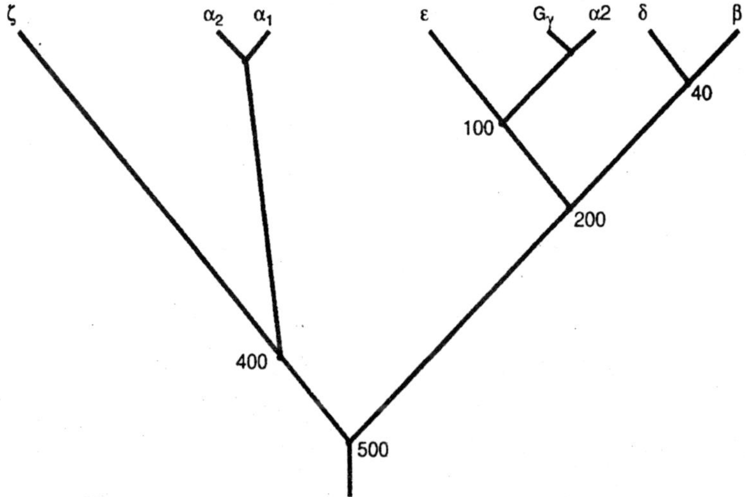

Fig. 14.3. A phylogeny of the human globin genes.

These inferred dates pose an interesting paradox. If we compare the sequences of the human α_1- and α_2-globins, the split apparently happened very recently. In fact, it may post-date the split between humans and the great apes and it certainly post-dates the split between the great apes and the rest of the primates. In this case, the primates outside the great apes should have only one α-globin. The prediction, however, is false: all the primates have two α-globin molecules. The full taxonomic distribution of the α-globins suggests that the gene duplicated at least before the origin of the mammals (about 85 million years ago), and probably before the split between mammals and birds (about 300 million years ago). Yet the similarity of the α_1- and α_2-globins in humans suggests the genes duplicated about 1 million years ago. Which of the figures is correct? This paradox, which is a common characteristic of evolution in gene families, is called *concerted evolution.*

Gene Family

If each tandem repeat of the ribosomal RNA gene evolved independently by mutation, drift, and selection, the different genes might gradually diverge from one another over time. Different mutations arise in different genes occasionally in the lineages of two species and become fixed. Because the evolutionary changes accumulate independently both in the different genes and in the different species, we might expect that the similarity between the genes in a gene family within a species would be approximately the same as the similarity between copies of the same gene in different species. In the modern species A and B at the top of Fig 14.4, if you examine one gene in the gene family, such as the gene at the extreme left, reveals one difference between it in the two species; there is also one difference between this gene and the fifth and seventh gene within species A.

This structure provides only a rough prediction of how evolution could proceed. The different genes in a gene family within a species have been separate since the duplication, whereas copies of the same gene in different species have been separate since the species split apart. If the duplication occurs in both species, then the speciation event will be more recent than the duplication. Other things being equal, the copies of the same gene in different species should then be more similar than the duplicates within a species. On the other hand, correlated selection pressures may operate within a species, tending to make the duplicated copies within a species more similar.

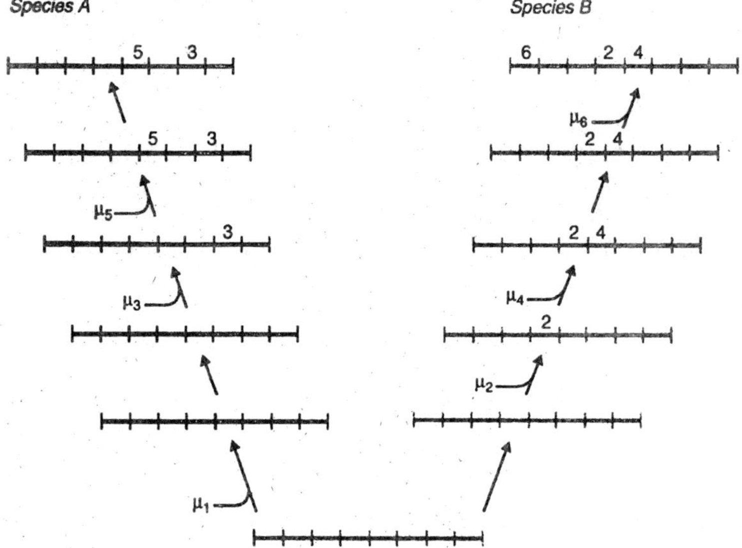

Fig. 14.4. Evolution of a gene family in two evolutionary lineages (in separate species).

And what about the facts? The first study, by Brown et al (1972), was made for the genes that code for ribosomal RNA in *Xenopus*. The 18S and 28S regions both fit the expected pattern. The sequences were almost identical in all tandem repeats of both *X. borealis* and *X. laevis*, suggesting strong stabilizing selection. For the non-transcribed spacer region (NTS), however, the pattern was different. Again, within a species the sequence was similar between tandem repeats, but this sequence differed in the two species. The NTS regions of *X. borealis* and *X. laevis* showed no more similarity than would two quite different genes—whereas within a species the NTS regions all had the same sequence. Thus, the NTS region showed concerted evolution.

The α-globin genes of primates described earlier can be used to illustrate the same principle. All primates, we saw, have two α-globins; we can therefore assume that the common ancestor of primates had two α-globin genes. The sequence of each α-globin gene differs between primate species. In the great apes, for example, any two species differ by about 2.5 amino acid substitutions in each gene. However, within a species, the α_1- and α_2-globins differ by only about one-tenth of that amount (this difference has been shown to be true not only for humans, but also for chimpanzees, gorillas, and rhesus monkeys). If one gene accumulates about 2.5 amino acid changes in the time between two species, then two different genes (α_1 and α_2) that have been separated for 300 million years should have accumulated many more changes—if they have been evolving independently. The conclusion is that they have not evolved independently; they have evolved in concert.

Why do gene families show concerted evolution? The concept is very difficult to explain if the genes evolve by mutations that occur independently at each gene locus. Indeed, independent mutations combined with neutral drift can be ruled out as an explanation. If the α_1- and α_2- genes drifted at random independently, they should be much more different than observation shown them to be within each species. Independent mutations and selection can, in principle, explain concerted

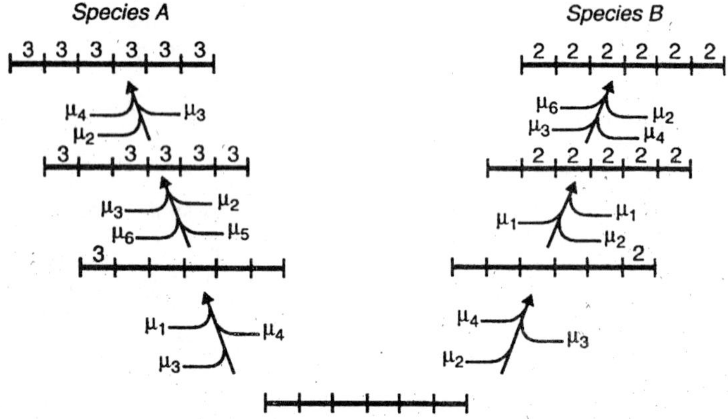

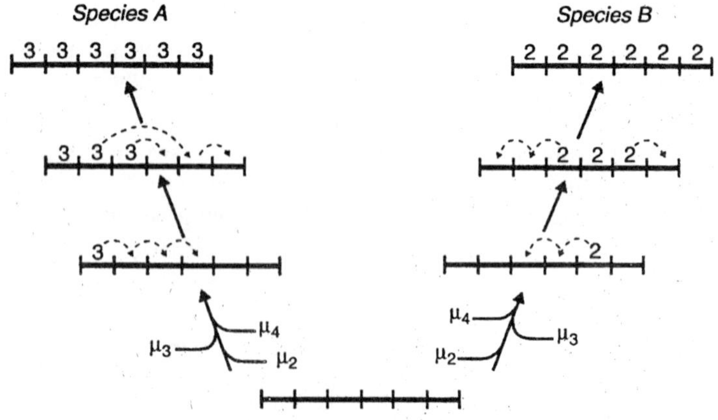

Fig. 14.5. Concerted evolution.

evolution. The same mutations would have to arise independently in both α_1- and α_2-globins in each species and then be fixed by selection, perhaps because one variant of the two genes was favoured in the conditions under which humans evolved, whereas another variant of both was favoured in the conditions under which chimpanzees evolved.

For the two globins in a number of primate species, this theory is a possibility. However, as the gene family showing concerted evolution grows larger, the explanation becomes increasingly incredible. When we reach the ribosomal RNA genes, with their innumerable tandem repeats, it is beyond belief that the same mutation could have occurred independently at each locus and been fixed by selection. In this case, it is generally accepted that some genetic mechanism must act to homogenize the different members of the gene family. One sequence somehow is copied horizontally from one gene to the others. Mutations are not occurring independently at the different loci.

Two main mechanisms have been suggested as causing concerted evolution: unequal crossing over and gene conversion. We encountered unequal crossing over when we discussed gene duplication. The same process generates chromosomes with different numbers and combinations of genes; these chromosomes can provide the mutational raw material for selection or drift to homogenize the gene family. Selection could directly favour more homogeneous gene clusters, or it could simply eliminate clusters with too many or too few copies of a type of gene.

In *gene conversion*, one of the alleles at a locus is converted into the other allele. Thus, when the heterozygote f_1f_2 segregated, instead of the Mendelian proportions of one f_1 for one f_2, two f_1 or two f_2 (and none of the other allele) would emerge. The same process could take place horizontally between the members of a gene family, with the variant at one locus being copied into the other locus. The mutational raw material for concerted evolution would again be produced.

Gene conversion could be either biased or unbiased. Biased gene conversion means that it favours one variant of the gene rather than another. For example, two f_1 genes might be produced more often than two f_2. Biased gene conversion is a kind of directed mutation, and it increases the chance that a gene cluster will be homogenized. Unbiased gene conversion means that the production of two f_1 or two f_2 genes is equally likely.

Unbiased gene conversion and unequal crossing over do not by themselves produced concerted evolution, any more than undirected mutation alone produces other types of evolution. They account only for the origin of variants—that is, they cause some individuals of a population to have more homogeneous gene families than other individuals. Selection and drift are still needed to explain how one of the gene family variants becomes fixed in the population. In the case of drift, a lateral "march to homozygosity" would occur between loci much like the process for a single locus.

The relative importance of unequal crossing over and gene conversion is unknown, by Hillis et al. (1991) found that the genes for ribosomal RNA showed concerted evolution even in an asexually reproducing lizard, which does not have crossing over. Gene conversion was therefore presumably responsible for the concerted evolution. Moreover, one sequence variant was favoured, suggesting that gene conversion may be biased in this case.

So far we have considered concerted evolution in a single gene cluster. The process is not confined to gene clusters on one chromosome, however. The ribosomal RNA genes in primates appear in clusters on five chromosomes, but the genes on different chromosomes show concerted evolution like genes on the same chromosome. This evolution is probably made possible when the regions of the chromosomes encoding the ribosomal RNA genes physically come into association. Unequal crossing over, or gene conversion, could then take place, just like that hypothesized to allow concerted evolution among genes on the same chromosome.

Concerted evolution is usually confined to only some of the genes in a gene family. For example, neither the α- and β-, nor the β- and γ-globins evolve in concert, even though they belong to the same gene family. The reason is that gene conversion takes place only between genes with some degree of sequence similarity. The chance of gene conversion is probably roughly proportional to the sequence similarity between two genes. A newly duplicated gene, therefore, might undergo one of two evolutionary courses. If gene conversion happens before the two genes have diverged dramatically, they may evolve in concert. If, however, they diverge too far for gene conversion to be possible, they will "escape" from one another, and their future evolution will happen independently. At a further stage, when they have diverged far enough, we should cease to classify these genes as members of the same gene family. Which of these two fates a pair of gene undergoes will be determined by the relative rates of gene conversion and of the mutations that break down sequence similarity. If the former is high relative to the latter, concerted evolution is likely; if the opposite is true, the genes will escape and diverge from one another.

The chance of gene conversion is probably not simply proportional to sequence similarity. The insertion of a mobile genetic element into a gene can prevent future conversion of that gene, or of part of it to one side of the insertion. Schimenti and Duncan have suggested that the insertion of an *Alu* element has prevented gene conversion, and concerted evolution, between certain genes in the β-globin families of cows and goats. The chance of concerted evolution is, therefore, controlled by both the sequence similarity of two genes and the insertion of a mobile genetic element in them.

In summary, the genes in gene clusters evolve in parallel. This concerted evolution is probably not caused by selection and drift operating on independent mutations at each locus; rather, it requires some mechanism for the same mutation to arise at the different loci. Gene conversion or unequal crossing over allow mutations to move between loci, and selection or drift could then fix the same mutation at all loci. In the case of biased gene conversion, the mutational mechanism itself will help to fix the favoured variant. The result would be the observed pattern of concerted evolution.

Coding of Genes by DNA

So far we have been concerned with genes that code for proteins, or at least for RNA. However, much of the DNA in the organisms of many species probably does not code for anything. Consider humans as an example. The human genome contains about 3×10^9 nucleotide pairs, and geneticists disagree about what proportion of it is *coding DNA* (i.e., DNA that is transcribed to produce proteins or regulate the production of proteins). Estimates are fallible, but the figure may be about 10-25%; the highest estimates go up to 50%. Even if 50% of the genome does code for genes, a large proportion of the genome still does something other than coding for, or controlling the production of, proteins. DNA that does not code for genes is called *non-coding* DNA.

DNA re-association experiments allow an estimate of the proportion of non-coding DNA. In these experiments, the DNA is first heated and melted into single strands, and the single-stranded DNA is then allowed, in a cooler solution, to join together (re-associate). The rate at which the strands re-associate is controlled by their sequence similarity. Early work identified three classes of DNA. A part of the DNA re-associated quickly; it is mainly highly repetitive DNA, made up of large numbers of repeats of simple sequences. At the other extreme, "single-copy" DNA joined together relatively slowly; this part of the DNA probably codes for most of the genes. In between is a class of middle-repetitive DNA, which is repeated but not as much as the highly repetitive DNA. On the assumption that the single-copy DNA serves as the coding part of the genome, we can use the proportion of single-copy DNA (i.e., slowly re-associating DNA) as a first estimate of the coding DNA in the genome.

Table 14.1. list the genome sizes and proportion of single-copy DNA for a number of species. These numbers suggest, in two ways, that a large quantity of non-coding DNA exist.

First, the figures for single-copy DNA are often much less than 100%. Only one-fifth of the DNA of the toad *Bufo bufo* is single-copy DNA, for example. If only single-copy DNA acts as the coding part, then four-fifths of *B. bufo*'s DNA is non-coding. This observation is generally true in eukaryotes, but not in prokaryotes. Bacterial DNA appears to be much more economically organized.

Second, large differences appear between species, and these differences can hardly all be due to variations in the numbers of coding genes. Fifteen times as many genes may be needed to build a human as to build a protozoan. In the words of Orgel and Crick, however, "it seems implausible that the number of radically different genes needed in a salamander is 20 times that in a man." The obvious deduction is that the differences derive mainly from non-coding DNA. Thus, a genome consists of more than simply genes.

The experiments provide a second clue about the nature of the non-coding DNA—namely, that it is often repetitive. Geneticists distinguish a number of kinds of repetitive DNA. A first distinction is between *tandem* and *scattered repeats*—that is, whether the repeat sequences appear next to one another (in tandem) or are scattered through the genome. Repetitive DNA can also be classified by the length and number of the repeat units.

No generally agreed classification exists for repetitive DNA, although the suggested classifications are all similar. We will use the following classification developed by Charlesworth et al. (1994). There may be other kinds of repetitive DNA that are not readily classifiable in one of these categories—or there may not be: the question is a topic of current research. (The word "locus" features in the descriptions below. It is used here in a slightly different way from traditional genetics: a locus for a tandem repeat is the site in the chromosome where a whole series of repeats is present. Thus, it refers to the whole array, not just one unit within it.)

Table 14.1. Amount of DNA percentage of single copy DNA in various animal species.

Invertebrates			*Vertebrates*		
Species	*1C (pg)*	*% sc DNA*	*Species*	*1C (pg)*	*% sc DNA*
Protozoa			Protochordata		
Tetrahymena pyriformis	0.2	90	*Ciona intestinalis*	0.2	70
Coelenterata			Pisces		
Aurelia aurita	0.7	70	*Scyliorhinus stellatus*	6.1	39
Nemertini (Rhyncocoela)			*Leuascus cephalus*	5.5	44
Cerebratulus	1.4	60	*Raja montagui*	3.4	47
Mollusca			*Rutilus rutilus*	4.8	54
Aplysia california	1.8	55	Amphibia		
Crassostrea virginica	0.7	60	*Necturus masculosus*	83.0	12
Spisula solidissima	1.2	75	*Bufo bufo*	7.0	20
Loligo loligo	2.8	75	*Triturus cristatus*	21.0	47
Arthropoda			*Xenopus laevis*	3.1	75
Prosimulium multidentatum	0.18	56	Reptilia		
Drosophila melanogaster	0.18	60	*Natrix natrix*	2.5	47
Limulus polyphemus	2.8	70	*Terrapene carolina*	4.1	54
Musca domestica	0.9	90	*Caiman crocodylus*	2.6	66
Chironomous tentans	0.21	95	*Python reticulatus*	1.7	71
Echinodermata			Aves		
Strongylocentrotus purpuratus	0.9	75	*Gallus domesticus*	1.2	80
			Mammalia		
			Homo sapiens	3.5	64
			Mus musculus	3.5	70

Tandem Repeats

1. *Microsatellites*. Repeats of short (2-5 base pairs) nucleotide sequences. The number of repeats varies between loci, but the average is on the order of approximately 100 repeats. Many microsatellite loci are scattered through the genome; an average human, for example, contains about 30,000 microsatellite loci. Microsatellites have also been detected in other vertebrates, insects, and plants.
2. *Minisatellites*. Repeats of longer (approximately 15 base pairs) nucleotide sequences. The number of repeats varies minisatellite loci, and many loci are scattered through the genome. The average length of any one locus is usually about 500-2000 nucleotides. Minisatellites have been studied in humans and other vertebrates, fungi, and plants.
3. *Satellite DNA*. The size of the repeated unit varies in different cases, with some being as small (5-15 base pairs) as micro- and minisatellites, while others are larger (about 100 base pairs). They are often found in large blocks of 1000 or more repeats of the unit sequence in regions of the chromosome near the centromere or the telomere. They do not seem to be as variable in the number of repeats per site as are micro- and minisatellites.

Scattered Repeats

Longer sequences (on the order of approximately 100 base pairs) that are distributed throughout the genome, usually in single copies bounded by other sequences rather than in tandem repeats. Three examples from humans are the sequences called *Alu*, *Kpn*, and poly (C-A); the three together make up about 20% of the human genome.

A unit sequence of any of the three kinds of tandem repetitive DNA can be found at more than one place in the genome. At all those sites they are found in the form of tandem repeats. Scattered repeats are also found at many sites, but usually with one a single copy of the sequence appearing at each site.

Two main evolutionary hypotheses have been suggested to explain the presence of this repetitive, non-coding DNA. One hypothesis says that such DNA is functional, even though it does not actually encode genes. It may be needed for some regulatory or structural reason, for example, or perhaps it keeps the genes apart or correctly configured in the DNA molecule's three-dimensioned shape. Alternatively, the repetitive DNA may be selfish DNA—either neutral "junk" DNA, or parasitic DNA.

We will concentrate on the "selfish DNA" hypothesis here, for two reasons. First, it is widely accepted as the explanation for much of the repetitive non-coding DNA. Second, it is a new evolutionary concept. If the repetitive DNA is functional, it will evolve just like genic DNA and requires no special discussion. If the DNA is non-functional for the organism that contains it, something different is occurring. It is also worth noting that comparisons of non-coding repetitive DNA made between different species have often shown the presence of concerted evolution, as in the gene families of coding DNA discussed earlier. The mechanisms presumably operate in much the same way in the two cases, and we will not discuss concerted evolution here.

Selfish DNA

In 1980, Doolittle and Sapienza, and Orgel and Crick, made more explicit an idea that many other geneticists had also pondered—the idea that some, or all, repetitive DNA may have no use for the organism. It may instead be *selfish DNA*. Selfish DNA is non-transcribed and non-coding, and contributes nothing to the well-being of the organism. In most cases, such DNA is selectively neutral except for the energetic burden of replicating it. If it was excised, the organism would suffer no disadvantage. After a stretch of selfish DNA originates, it is passively replicated and passed on from parent to offspring. Changes in its frequency in the population would occur by drift. If a non-coding sequence was not neutral—perhaps because it interfered with the construction or metabolism of the organism—it would be selected against. Likewise, if it accumulated to such an extent that the cell cycle was slowed by the need to replicate the entire DNA sequence, selection would probably act to reduce it. Provided the quantity of a particular sequence is not excessive, is not transcribed, and accumulates in parts of the genome where is does not interfere with genic regulation and transcription, there is no reason why selfish DNA should not evolve.

Selfish DNA might be either "active" or "passive". The sequence itself might not influence the chance that it spreads in the DNA and is retained. It is then a passive kind of selfish DNA and could accumulate as "junk DNA" in the genome. Alternatively, a particular sequence might, for some reason related to its sequence, have a better than average chance of spreading through the DNA. These sequences would be a more active, parasitic kind of selfish DNA, and they would proliferate until checked by natural selection because they interfered with vital DNA functions or became so abundant that their replication imposed a significant cost on the organism. For the possible examples of selfish DNA we shall discuss here, we do not know for certain to which class they belong. This distinction should be kept in mind, however.

How might selfish DNA originate? In theory, any kind of mutation could give rise to selfish DNA. A point mutation could inactivate a gene, converting it into a pseudogene. If the inactivated gene was redundant and the new pseudogene was not in a position to interfere with any important function, the mutation could be fixed by drift and the pseudogene retained as passive selfish DNA. For repetitive DNA, other kinds of mutations would be needed.

We will discuss tandem and scattered repetitive DNA separately, because they are believed to originate by different mechanisms. In this section, we will examine three ways of creating tandem **repetitive** DNA briefly.

The three processes that have been suggested to create tandem repeats are called slippage, unequal crossing over, and rolling circle replication.

Slippage occurs when one DNA strand is being created by replication from another strand. The two strands may slip relative to one another by a few nucleotides, such that either a few nucleotides

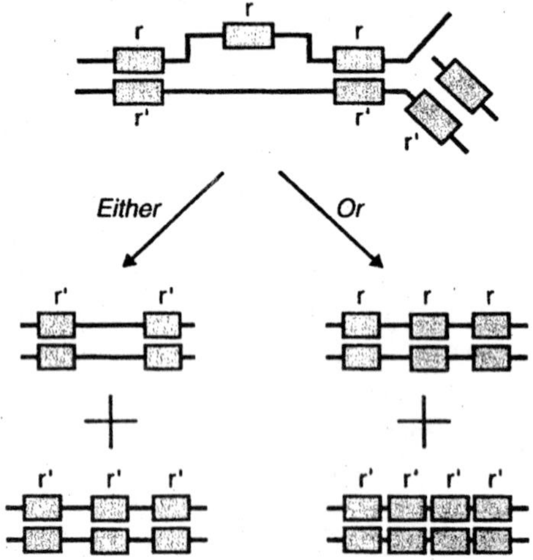

Fig. 14.6. Slippage occurs when a sequence of DNA is copied twice, generating a new strand with an extra copy of that sequence, or missed out generating a new strand with one copy less.

are missed or copied twice. Such an event is most probable in a region of short repeats, and the result is a new strand that has either more or less repeats than the original strand. Slippage occurs only over short distances, and is the most likely mechanism by which new length variants arise for microsatellite sequences. It may produce new length variants in minisatellites as well, although unequal crossing over is also important in this case.

Unequal crossing over can produce a strand with a greater or lesser number of repeats, and successive rounds of unequal crossing over can build up or break down long sequences of repeats. Unequal crossing over has not yet been confirmed to operate in minisatellites, but it is strongly suspected to play a role based on indirect evidence. While slippage could operate in the short unit satellites, unequal crossing over could operate in all of them.

The third mechanism, *rolling circle replication*, has also been suggested to affect all types of satellites. This method of replication has been well studied in bacterial plasmids and some viruses; the idea that it operates in eukaryotes is more speculative. It would require the excision of a sequence from the chromosome. The sequence would then have to form a circle, after which it could multiply by rolling circle replication. The replicated sequences would then have to be reinserted into the chromosome.

All of these processes are mechanisms for the origin of new variants. When they operate, one individual in the population will have a new length variant at a site of non-coding repetitive DNA. Drift or selection will then be needed for the variant to spread through the population. In the models of evolution, evolution occurred when rare variants had their frequencies altered by selection or drift. Slippage, unequal crossing over, and rolling circle replication are all analogous to the "mutation" phase of those earlier models, and are thus only one part of an evolutionary explanation.

Under both slippage and unequal crossing over, new length variants of a unit sequence become more likely to arise if that sequence is present in more lateral copies in the genome. If only one copy of a sequence exists it is unlikely to misalign. If several repeats of the sequence are present, misalignment is more probable, as copy two on one strand aligns with copy three on the other, or copy eight on one strand aligns with copy six on the other. The more copies of a sequence that exist, the more likely this misalignment becomes. If, therefore, a genome contained two sorts of neutral, functionless sequence—one in single copy and the other in multiple repeats—then (other things being equal) the repeated sequence would be more likely to grow laterally through the genome during evolution. If the highly repetitive rows of tandem repeats are actually selfish DNA, their lengths will fluctuate through evolutionary time. Unequal crossing over, slippage, and drift will constantly be causing them to shrink and grow. At any one time, some very long regions of repeats will be found, and these regions will be recognized as highly repetitive DNA. After a few million years, a currently long repeat could potentially have shrunk and some other, previously shorter series lengthened. The frequency distribution of lengths will likely remain probabilistically constant.

Minisatellites (Short Repeats)

While studying the human myoglobin gene in the early 1980s, Jeffreys discovered a short sequence of repeated DNA within an intron. He used the short sequence as a "probe" to see whether the same sequence was present anywhere else in the genome. The probe hybridized at several regions. Jeffreys then extracted the DNA from all these regions, and found that each DNA sample consisted of a unit sequence in a row of tandem repeats. The number of repeats at each site was highly variable between individuals; one site, for example, might contain 25 repeats in one individual, 5 in another individual, and 47 in another. Although the unit sequence itself is variable, a common core sequence has been characterized and is 16 (or perhaps 8) bases long. The particular sequence is not the only sequence that can form this sort of scattered short set of repeats, but it is the most studied. Jeffreys called these sequences minisatellites; the name (like

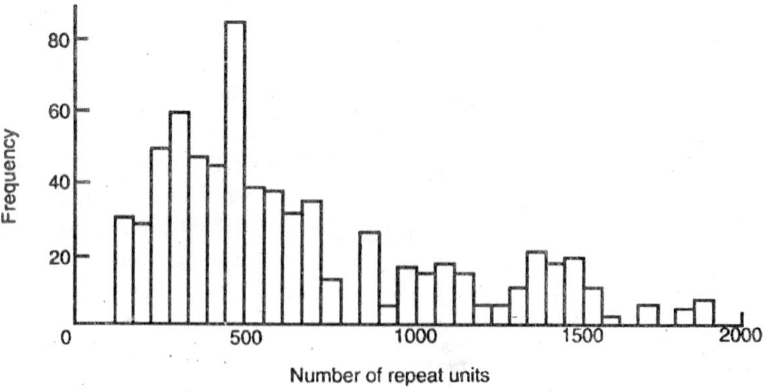

Fig. 14.7. Frequency distribution of numbers of repeats at one minisatellite locus in a total of 344 individuals (688 gametes).

that of microsatellites and satellites) derives from the way they can be detected when DNA is centrifuged.

Why do minisatellites vary so much in the number of repeats? We saw earlier that slippage and unequal crossing over are thought to generate new length variants, and we can now look at two predictions of this idea. Both concern the heterozygosities of different minisatellite loci. The alleles at a minisatellite locus are the different length variants, which differ in their number of repeats of the unit sequence. We can measure the frequencies of the different length variants at each locus, and calculate the heterozygosity for each by the standard formula.

The first relation to consider exists between the heterozygosity of a locus and the similarity of the unit sequence. So far we have described repetitive DNA as if the unit sequence were identical in all repeats. In practice the sequence can differ slightly in terms repeats, with greater differences appearing at some loci. When Stephan and Cho plotted the heterozygosity against the sequence similarity for 10 minisatellite loci in humans, they found a strong negative relation. The loci that are most variable in numbers of length variants possess the most similar unit sequences. The relation is expected if new variants arise by unequal crossing over, for two reasons. First, unequal crossing over is most likely if the unit sequences are more similar. Second, unequal crossing over causes the same variant to be spread through the DNA at a site. If unequal crossing over has happened more frequently at one locus than at another, the unit sequences should look more similar at those loci where the process is more common. If unequal crossing over has not occurred much at a locus, the sequences of the repeats will have diverged more.

Fig. 14.8b shows another relation with heterozygosity for different human minisatellite loci. The more heterozygous loci have higher mutation rates. The mutations in this case form new length variants. If a locus is fixed for one length variant, unequal exchange is unlikely to operate. The more length variants present at a locus, the more likely it is that two different ones will cross over unequally and produce a new variant. That is, the more variety introduced into the system, the more variety it is likely to generate. If the mutations arise by a more conventional genetic mechanism, there would be no reason to predict this trend; if they arise by unequal crossing over, we would expect this result. The mutation rates were estimated by tracing minisatellites through human pedigrees, to see when new variants arose. Notice that the rate is high, up to 5% per gamete.

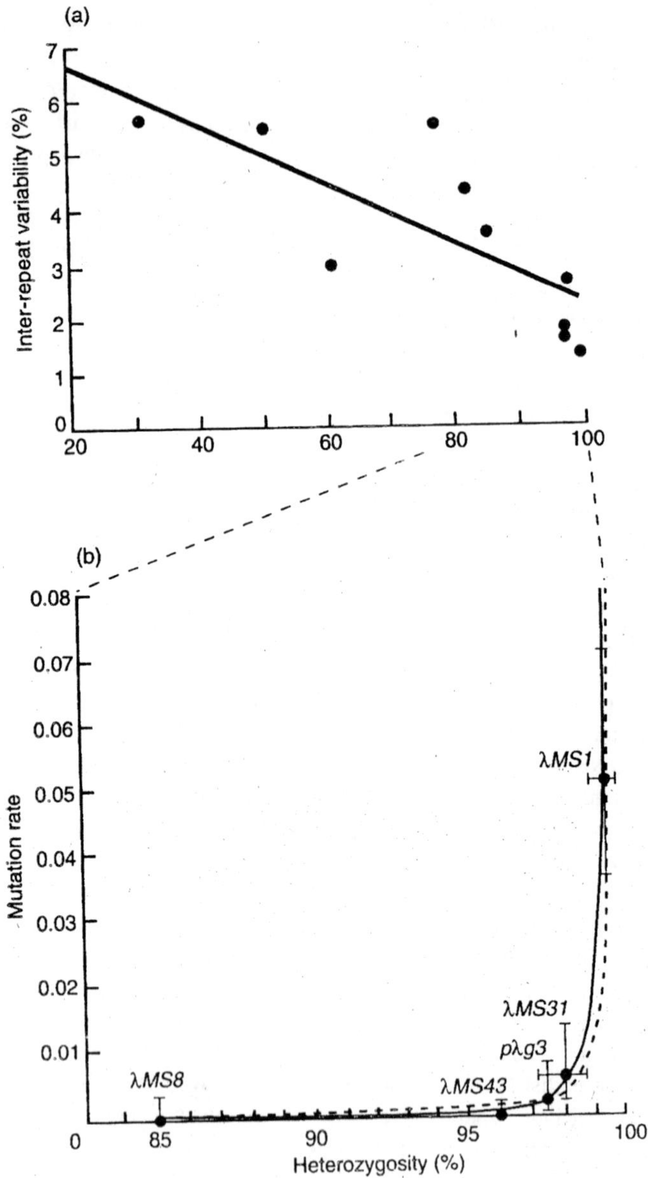

Fig. 14.8. Two relations of the variation of different minisatellite loci in humans. (a) The sequences of the units are more similar at loci that are more heterozygous. (b) Mutation rates are higher at more variable loci. Both relationships make sense if length variants at minisatellites mainly originate by unequal crossing over.

The high mutation rates of minisatellites allow them to be used in *genetic fingerprinting*. New variants arise at a high enough rate for every individual (except monozygous twins) to have a unique frequency distribution (or "profile") of minisatellites. Genetic fingerprints are more forensically useful than real fingerprints. Not only is an individual's genetic fingerprint unique and identifiable from small quantities of body substances, but it is also heritable. The profiles of a father and his children, for instance, are more similar than two random members of the same population. Genetic fingerprinting can, therefore, be used in paternity testing as well as in straightforward identification. Minisatellites similar to those in humans have been found in many other species, and the genetic fingerprinting probe has become an important method of tracing paternity in behavioural ecology. Microsatellites are also used in genetic fingerprinting.

Scattered Repeats

Scattered repetitive DNA is thought to originate mainly by *transposition*. Certain sequences of DNA can, under appropriate circumstances, copy themselves elsewhere in the genome. These mobile genetic sequences are called transposable elements, or (more informally) jumping genes. The molecular biology is complex, but we can distinguish two types of transposition; these processes differ at a molecular level according to whether an RNA intermediate is present. The difference is evolutionarily interesting because transposition by an RNA intermediate may be more likely to increase the number of copies of the sequence in the DNA.

We can make a second distinction between *replicative* and *conservative* transposition. In Fig. 14.9a, two copies of the sequence exist after the transposition, whereas only one was found before transposition. In Fig. 14.9b, however, only one copy exists both before and after transposition. (It is possible for transposition by DNA intermediate to be replicative. For instance, the excised sequence may jump from one of the two strands produced after the DNA has replicated itself in the cell cycle, to a position upstream of the replication fork, where the DNA has yet to be replicated.)

Replicative transposition could theoretically produce a scattered repeat sequence. After its origin in one individual, it could then increase in frequency (by selection or drift) to result in the pattern of scattered repeats we now observe. Did real scattered sequences evolve in this way? The best evidence that a scattered repeat sequence had originated by transposition would be found by observing the sequence transposing.

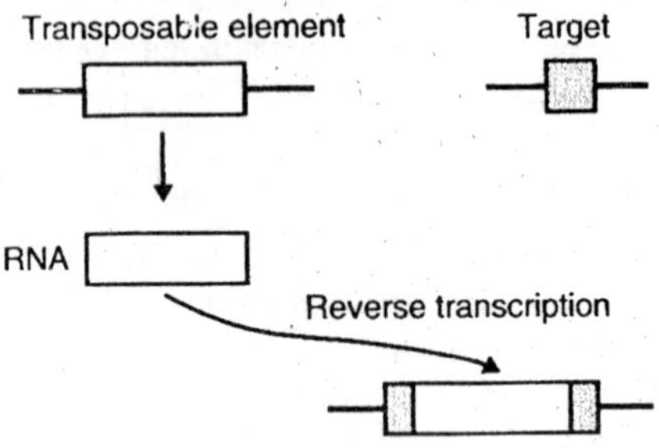

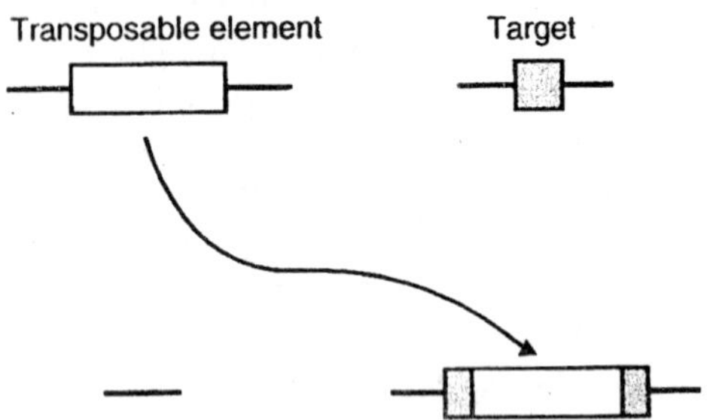

Fig. 14.9. Two main kinds of transposable elements. (a) Retroelements, which transpose by the reverse transcription of an RNA intermediate. (b) Transposable elements, which transpose by excising from one site and inserting elsewhere, without an DNA intermediate.

A more general method is to study the characteristics of the DNA sequences that are known to transpose, and of the scattered repeats, to see whether the latter resemble the former. Transposable elements have some characteristics features, and the similarity between these and at least some well-studied scattered repeat sequences is enough to make a compelling case for their origin by transposition. Let us look first at some known (or strongly suspected) transposable elements, and then at an example of a scattered repeat.

The simplest transposable elements lack the enzyme reverse transcriptase. They possess only a "transposase" enzyme together with one or more other genes, and are transposed by a DNA intermediate. Such elements usually have inverted repeat sequences at either end; the same sequence is found at both ends, with one sequence forming a mirror image of the other. The most famous examples are the first transposable elements to have been described—the ones in maize discovered by McClintock. Other transposable elements have the gene for reverse transcriptase as well, and they are classified as retroelements. Some retroelements can live only in the genome; the element called *copia* in the fruitfly *Drosophila* is an example of this phenomenon. Other retroelements can live independently or within a host genome; these elements include the retroviruses, such as HIV, the agent of AIDS.

Which sequences in our DNA may have been produced from these kinds of transposable elements? Transposition has been most studied in bacteria, and our knowledge for multicellular organisms is less certain, but some likely examples have been identified. The *Alu* sequence is a hot candidate in mammals. *Alu* is the name for a characteristic sequence recognized by the restriction endonuclease *Alu* 1. The sequence is about 281 bases long and is found throughout the human genome; each genome contains about half a million copies. About 5% of human DNA consists of the *Alu* sequence. Few, if any, of the sequences are transcribed, and the *Alu* sequence may have no function.

Where does the *Alu* sequence originate? One favoured hypothesis is that it ultimately comes from the gene for a functional ribosomal RNA molecule called 7SL RNA. The gene for the 7SL RNA could have first been picked up by a retroelement and copied back into the DNA by reverse transcription; the many *Alu* sequences are believed to have been formed as second-order derivatives of the 7SL RNA gene, by recurrent rounds of retroelemental reverse transcription from a number of reverse-transcribed copies of the gene itself. The *Alu* sequence itself is not strictly a retroelement because it lacks the gene for reverse transcription, but it looks like a sequence that is derived from reverse transcription. Perhaps it is picked up by a retroelement, and the gene for reverse transcription is lost when it is copied back into the DNA. Although the sequence has not been proved to be functionless, a strong suspicion exists that it is an example of selfish DNA. The large numbers of copies of *Alu* suggest that something about its sequence makes *Alu* particularly likely to be copied; it would then be an example of active selfish DNA.

Transposable elements can have interesting evolutionary consequences, even if they evolved originally by neutral drift. The insertion of an *Alu* sequence can prevent gene conversion between a pair of genes. Concerted evolution between the two genes then becomes impossible. As a consequence, duplicated genes in which an *Alu* sequence is inserted would be more likely to show independent evolution than genes without such an insertion. A second suggested consequence involves chromosomal evolution. Chromosomes may be particularly likely to rearrange around *Alu* sequences, and the numbers and positions of the sequence may, therefore, influence the rate and form of chromosomal evolution. Finally, at the level of the DNA sequence, when a transposon inserts into a new gene, it causes a mutation. The recipient gene has a new sequence after receiving the transposon, which may (or may not) cause a mutation at the phenotypic level. McClintock first showed that mutations can be caused by transposition in maize. Transposition probably accounts for a substantial proportion of observed mutations. It may differ from other kinds of mutation in that it may occur more at some sites in the genome than at others. Mutations caused by nucleotide changes may be more randomly distributed.

C-Factor Paradox

Let us return to the observation that more DNA is present in the cells of some species, particularly eukaryotes, than is needed to code for their genes. The degree of excess is controversial, but no one doubts that at least some excess DNA exists in at least some species. (Likewise, no one doubts that at least some of the species differences in DNA content, such as between humans and bacteria, are due to differences in phenotypic complexity.) The apparent excess of DNA is sometimes called the C-factor paradox. What purpose does the excess DNA serve?

Selfish DNA is an attractive, if hypothetical, solution to this question. The excess DNA would then have no use or function, but would be selectively neutral, and passively replicated from generation to generation. The assumptions of this hypothesis are plausible—unequal crossing over, gene conversion, transposition, and neutral drift all do happen. Moreover, if the hypothesis is correct, it would explain why this apparently functionless DNA actually is functionless. It could also explain why the degree of excess DNA differs so much among species, because the amount of repetitive DNA will fluctuate through time as it evolutionarily grows by mutation and drift (or selection) and shrinks when selection acts against excessive accumulations of repeats. Under this scenario, different species would represent the still-frames in a moving picture. While this idea is simply a hypothesis, it is, however, a theoretically attractive hypothesis. At the present time, the best available explanation for the apparent excess of DNA in our cells is that it results from the passive accumulation of selfish DNA.

INDEX

C

D

E